Placer Gold Deposits
of the
Sierra Nevada

COMPILED AND EDITED BY

Paul D. Morrison

Published by Gem Guides Book Company • 1997

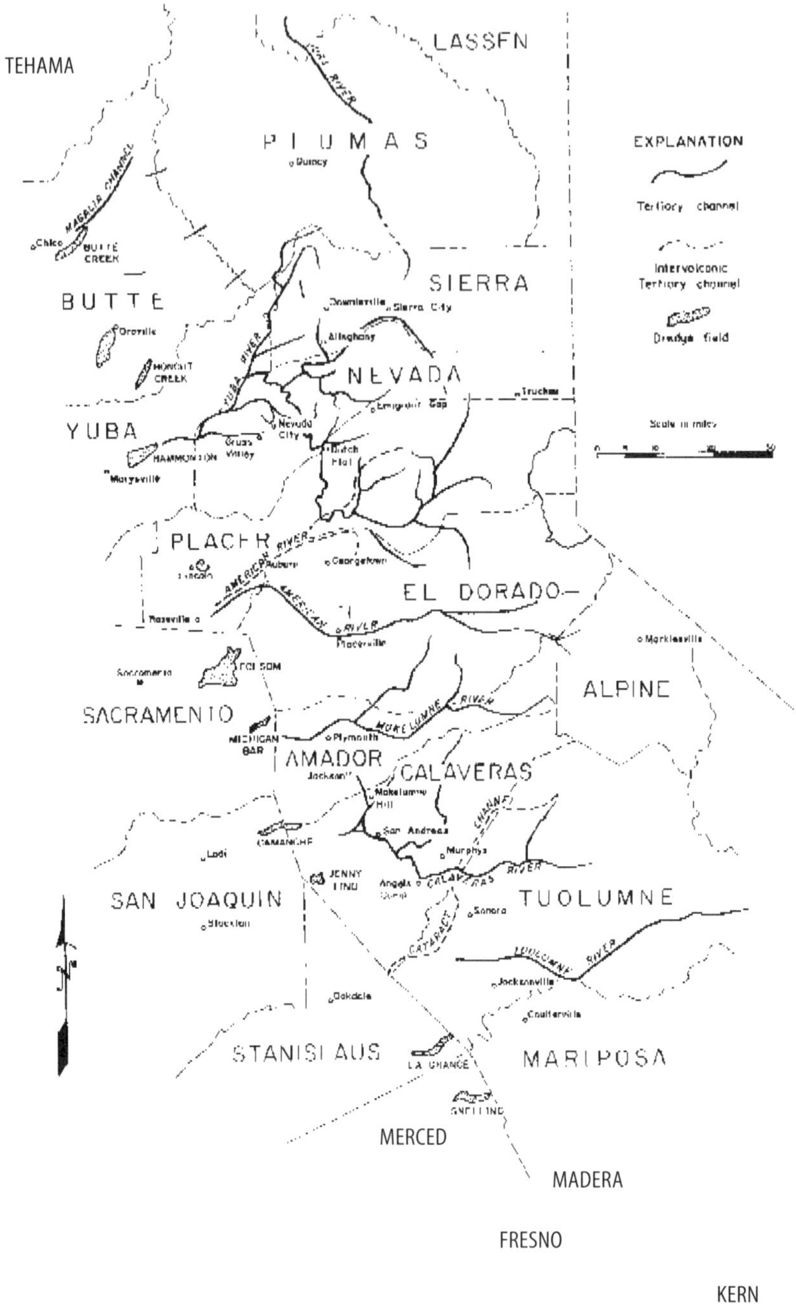

Published by Gem Guides Book Company
315 Cloverleaf Dr., Suite F
Baldwin Park, CA 91706

Cover design, book typography, and cartography by Art Waxer Designs, Berkeley, CA 94712.
Cover illustration: woodcut by H. Glintenkamp, 1930.
ISBN 0-935182-97-7.
Library of Congress Catalog Card Number: 97-60923
Manufactured in the United States of America

EDITOR'S PREFACE

This book was compiled from several sources, all of them government documents in the public domain, long out of print and accessible in some cases only in non-circulating collections at state or university libraries. Unless otherwise noted, the text is reproduced from the following volume:

Clark, William B. *Gold Districts of California.* California Division of Mines and Geology Bulletin 193. San Francisco: California Division of Mines and Geology, 1970. Manuscript submitted for publication 1963. Some revisions through 1969.

Other publications which have been excerpted in this book:

[1] Averill, Charles Volney. *Placer Mining for Gold in California.* San Francisco: State of California Division of Mines, October, 1946.

[2] Aubury, Lewis E. *Gold Dredging in California.* California State Mining Bureau Bulletin No. 57. San Francisco: California State Mining Bureau, 1910.

[3] Bradley, Walter C., and G. Chester Brown, F. L. Lowell, R. P. McLaughlin. *Mines and Mineral Resources of the Counties of Fresno, Kern, Kings, Madera, Mariposa, Merced, San Joaquin, Stanislaus.* San Francisco: California State Mining Bureau, 1915. Chapters of the State Mineralogist's Report, Biennial Period 1913-1914.

[4] Chandra, Deb K. *Geology and Mineral Deposits of the Colfax and Foresthill Quadrangles.* San Francisco: California Division of Mines and Geology, 1961. Special Report No. 67.

[5] Clark, William B., and Philip A. Lydon. *Mines and Mineral Resources of Calaveras County, California.* San Francisco: California Division of Mines and Geology, 1962. County Report No. 2.

[6] Lindgren, Waldemar. *Tertiary Gravels of the Sierra Nevada of California.* Washington, D.C.: Department of the Interior, 1911. United States Geological Survey Professional Paper No. 73.

[7] Julihn, C. E., and F. W. Horton. *Mines of the Southern Mother Lode Region.* Washington, D.C.: United States Department of the Interior, Bureau of Mines, Mineral Industries Survey of the United States. Bulletin No. 413, Calaveras County, 1938. Bulletin No. 424, Tuolumne and Mariposa Counties, 1940.

[8] MacBoyle, Errol. *Mines and Mineral Resources of Nevada County.* Sacramento: California State Mining Bureau, 1919. Chapters of the State Mineralogist's Report, Biennial Period 1917-18.

[9] ————. *Mines and Mineral Resources of Plumas County.* Sacramento: California State Mining Bureau, 1920. Chapters of the State Mineralogist's Report, Biennial Period 1917-18.

[10] ————. *Mines and Mineral Resources of Sierra County.* Sacramento: California State Mining Bureau, 1920. Chapters of the State Mineralogist's Report, Biennial Period 1917-18.

[11] Troxel, Bennie W., and Paul K. Morton. *Mines and Mineral Resources of Kern County, California.* San Francisco: California Division of Mines and Geology, 1962. County Report No. 1.

[12] Tucker, W. B. *Mines and Mineral Resources of Amador County, Calaveras County, Tuolumne County.* San Francisco: California State Mining Bureau, 1915. Chapters of State Mineralogist's Report, Biennial Period 1913-1914.

Where material from another source has been inserted into the text from Clark, the excerpt is preceded by a bullet [•] and concluded with a bracketed number corresponding to the above list.

No attempt has been made to include every district, claim or operator; those omitted from the present compilation generally lacked specific location information. The most complete lists of mines and prospects can be found in state mineral reports published in the *California Journal of Mines and Geology.* [5] above includes 21 quarto pages of detailed specifics of placer mines in Calaveras County alone. [6] has much additional description of the geology of the tertiary streams of the Sierra Nevada. Also recommended is:

Wilkerson, Gregg, and David Lawler. *Roadside Geology and Mining History of the Mother Lode.* Bakersfield: Bureau of Land Management Bakersfield District Office, 1994.

A bibliography, as originally published in Clark, is reproduced on pages 184-85.

The maps included here are intended as a general reference to obtaining more detailed state, county, U.S. Forest Service, or U.S.G.S. topographical maps, and should not be relied upon for guidance in the field. Several tracts shown may now be inaccessible due to building developments or water impoundments.

CONTENTS

MAPS

AMADOR COUNTY

Camanche-Lancha Plana

Location. This district extends along the Mokelumne River from the vicinity of Lancha Plana, Amador County, west through the Camanche-Wallace area, Calaveras County, to Clements, San Joaquin County, a distance of about 12 miles.

History. The creeks were first worked during the early part of the gold rush, and hydraulic mining of the terrace gravels followed. The town of Camanche was named for a town in Iowa. Later, the Chinese mined the river and reworked the old tailings. From 1904 to 1923, the river was dredged on a large scale by the American Dredging Company. Dragline dredging was done during the 1930s and bucket-line dredging from then until 1951. Much of the region was recently inundated by the Camanche reservoir. The output from dredging is estimated to be about $10 million.

Geology. The deposits consist of unconsolidated gravels in and adjacent to the Mokelumne River and floodplain deposits. Tightly packed shore gravels of Eocene age are found near Wallace. In the west portion of the dredging field, the gravels ranged from six to 50 feet deep while, in the east, they were six to 25 feet deep. Bedrock is clay and volcanic rock. Values recovered by dredging ranged from 10 to 25 cents per yard. The gold was fine grained and 850 to 900 in fineness.

Operations. American Dredging Co., 1904-23, three bucket-lines; Camanche Placers Ltd., 1935-, one bucket-line; Gold Gravel Products, 1935; Gold Hill Dredging Co., 1936-51, two bucket-lines; Lancha Plana Gold Dredging Co., 1926-40, one? bucket-line; Wallace Dredging Co., 1935-40, two bucket-lines.

Fiddletown

Location. This district is in northwestern Amador County in the general vicinity of the old mining town of Fiddletown, six miles east of Plymouth. It also has been known as the Oleta district.

History. Fiddletown was settled in 1849, reportedly by Missouri miners addicted to "fiddling." The district flourished in the 1850s when the drift and hydraulic mines were active. Attention was drawn to the area by Bret Harte's short story, *An Episode at Fiddletown.* The name proved offensive to one of the distinguished residents and, in 1878, he succeeded in having the name changed by legislative enactment to Oleta. In 1937, the California Historical Society, with the approval of local residents, obtained the restoration of the original name. There was some drift mining and dragline dredging in the district in the 1930s and early 1940s.

Geology. Numerous patches of quartzitic gravels remain that were deposited by several channels of the Tertiary Cosumnes River. One channel comes in from the Coyoteville area from the northeast and another from Volcano from the southeast. Some of the gravels are capped by andesite. Bedrock is chiefly graphitic slate, metachert and schist of the Calaveras Formation (Carboniferous to Permian). Limestone lies to the east and granodiorite to the north. There are a few narrow gold-quartz veins in the district.

Irish Hill

Location. The Irish Hill district is in northwestern Amador County about five miles north of Ione. It includes the Muletown and the Forest Home areas.

Geology. Several extensive patches of Eocene quartz-rich channel gravels and younger gravels exist. They were first mined by ground sluicing and hydraulicking and later by drag-line dredging. Bedrock consists of slate, phyllite, greenstone, and amphibolite. There are also several copper mines here.

Jackson-Plymouth

Location. A 20-mile-long belt of gold mineralization runs through western Amador County. On the belt, a portion of the Mother Lode, are the towns of Jackson, Sutter Creek, Amador City, Drytown, and Plymouth. Because of the uniform nature of the gold mineralization along this belt, the several districts-and sub-districts have been grouped together here under a Jackson-Plymouth heading.

History. This entire belt was settled early in the gold rush when the streams were placer-mined. Jackson was settled by California Spanish at least as early as 1849. It was first known as Botilleas, but the name was soon changed in honor of Colonel Alden Jackson. Sutter Creek was named for Captain John A. Sutter who visited the region in 1846. Amador City was settled in 1851 and Plymouth in 1852. Drytown flourished from 1848 until 1857 when rich placer deposits were worked. Most of the important lode deposits were discovered during the 1850s. The Argonaut mine was first developed in 1850, the South Spring Hill and Lincoln in 1851, the Plymouth in 1852, the Original Amador and Keystone in 1853, the Central Eureka in 1855 and the Kennedy in 1856. Lode mining developed into a major industry that was to last 90 years.

By 1875 mines such as the Keystone, South Spring Hill, Oneida, Old Eureka, and Plymouth had become large and highly profitable operations. However, the Argonaut, Kennedy, Central Eureka, Bunker Hill, Fremont-Gover, and Lincoln Cons. (Lincoln, Wildman, and Mahoney), major gold sources of a later date, did not become important until the 1880s and 1890s. The properties constituting the Plymouth Cons. mine were consolidated in 1883, the Kennedy Mining and Milling Company was organized in 1885, and the Argonaut Mining Company in 1893. From the 1890s until 1942, this belt was one of the more important gold-mining districts in the nation. The value of production ranged from $2 million to $4 million annually. Several thousand miners were employed, many of whom were of Italian, Austrian, and Serbian extraction.

All of the mines were shut down soon after the beginning of World War II. The Central Eureka mine was reopened in 1945, but because of greatly increased costs it was shut down again in 1953. This was the last active major gold mine on the Mother Lode.

Jackson-Plymouth was the most productive district of the Mother Lode belt, with a total output estimated by the author at about $180 million. If large-scale gold mining were ever to be done here again, it would be most desirable to consolidate the major mines and operate them as a unit. Some are connected underground and all produced considerable water.

The remaining surface plant of the Kennedy mine is now a museum. The large wooden tailing wheels and the superintendent's office at this mine, long noted landmarks, are historical displays. Several other distinctive old mine buildings have been preserved, including the Keystone mine office, which is a motel, and the Zeila office, which is a private home.

Geology. The gold deposits are in a north- and northwest-trending mile-wide belt of gray

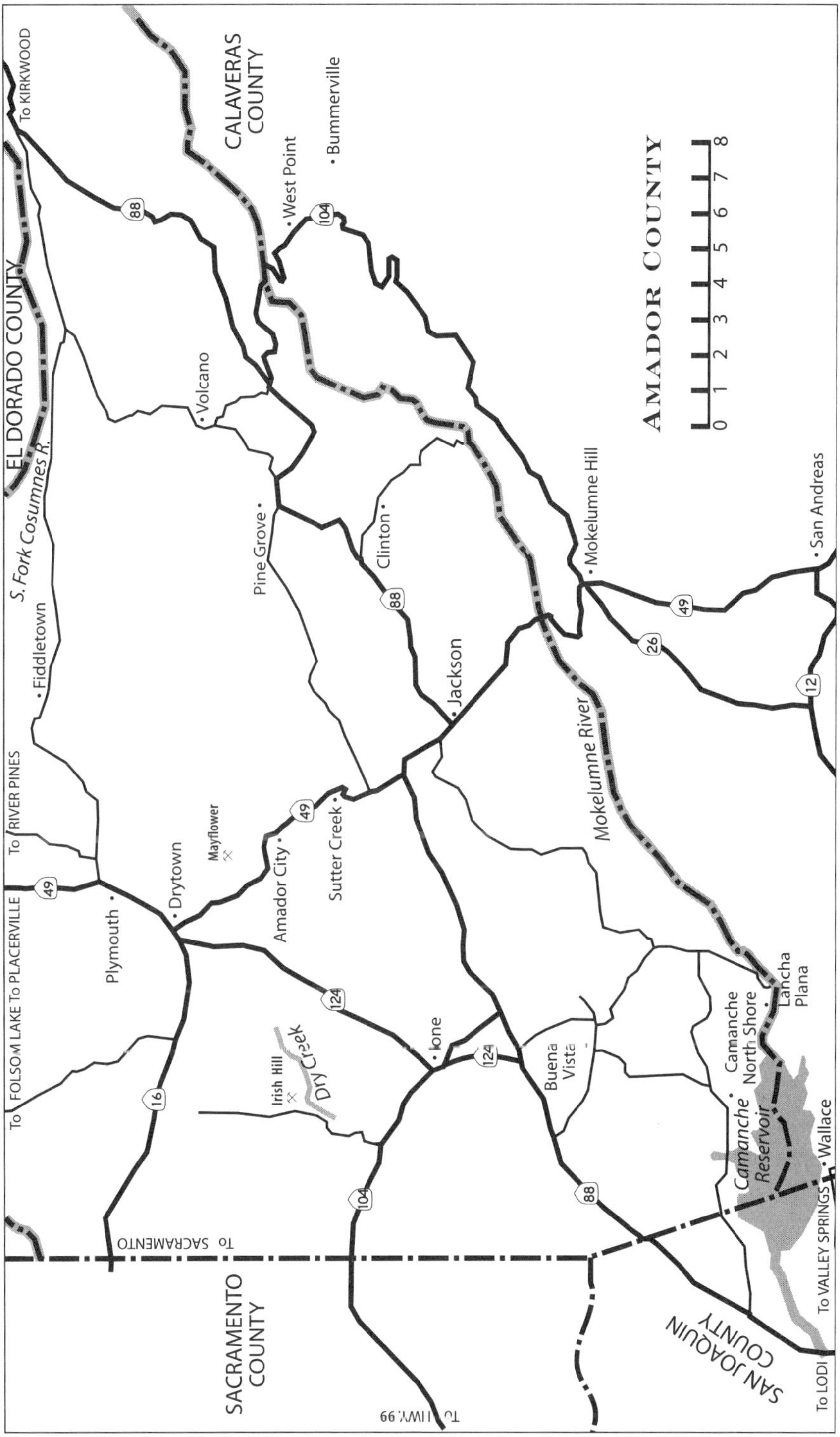
AMADOR COUNTY
0 1 2 3 4 5 6 7 8
EL DORADO COUNTY
CALAVERAS COUNTY
SACRAMENTO COUNTY
SAN JOAQUIN COUNTY
To KIRKWOOD
To RIVER PINES
To FOLSOM LAKE
To PLACERVILLE
To SACRAMENTO
To HWY. 99
To LODI
To VALLEY SPRINGS
S. Fork Cosumnes R.
Mokelumne River
Dry Creek
Camanche Reservoir
West Point
Bummerville
Volcano
Pine Grove
Clinton
Fiddletown
Jackson
Mokelumne Hill
San Andreas
Plymouth
Drytown
Mayflower
Amador City
Sutter Creek
Irish Hill
Ione
Buena Vista
Camanche North Shore
Lancha Plana
Wallace
88
104
49
26
12
124
16

© Gem Guides

to black slate of the Mariposa Formation (Upper Jurassic), with some interbedded and occasionally sheared conglomerate and minor sandy and gritty layers. Massive greenstone of the Logtown Ridge Formation (Upper Jurassic) lies west of the belt of Mariposa Formation slate. Metasedimentary rocks, chiefly graphitic schist, metachert and amphibolite schist of the Calaveras Formation (Carboniferous to Permian) are to the east. Several deposits of Tertiary auriferous channel gravels are exposed south of Jackson.

Mines. Alma, Alpine, Amador Gold $100,000+, Amador King $100,000+, Amador Queen No. I $100,000+, Amador Queen No. 2 $100,000+, Amador Star $100,000+, Anita, Argonaut $25.2 million, Ballard, Bay State $100,000+, Bellwether, Bunker Hill $5.1 million, Central Eureka Group $36 million, Crown Point, Detert Group, Fremont-Gover $5 million, Good Hope, Italian $140,000+, Hardenbergh $100,000+, Kennedy $34.28 million, Keystone $24 million, Lincoln $2.2 million, Mammoth, Mayflower, Mineral Point, Moore $564,000+, New London, North Star et al., Oneida $2.5 million+, Original Amador $3.5 million, Plymouth Cons. $13.5 Million, Potosi, South Eureka $5.3 million, South Jackson, South Spring Hill $1.1 million, Treasure $1 million, Valparaiso $100,000+, Wildman-Mahoney $5 million, Zeila $5 million+.

Pine Grove

Location. Pine Grove is in east-central Amador County about 10 miles east of Jackson. It includes the Irishtown and Clinton areas. It is a small Sierran east gold belt district that was first worked during the gold rush and has been intermittently prospected ever since.

Geology. The district is underlain chiefly by graphitic slate, schist, and metachert. Granodiorite lies to the east and west. A number of narrow quartz veins containing small ore shoots are rich in places. The ore commonly contains abundant sulfides, especially galena and chalcopyrite. A few small patches of Tertiary auriferous channel gravels overlie the bedrock.

Mines. Black Wonder, Contini, Mikado, Peterson, Pine Grove Unit, Rainbow, Red Hill.

Volcano

Location. This district, in north-central Amador County, is in the vicinity of the old mining town of Volcano, 15 miles northeast of Jackson.

History. The creeks were first mined during the gold rush, and the town was settled by soldiers from Stevenson's 1st New York Volunteer Regiment. The town received its name in 1850, as it was believed then that the limestone caves were related to a volcano. During the 1850s this was one of the richer placer mining districts in the state. Since the 1930s there has been a small output from dragline dredging in some of the creeks, and some of the channel gravels have been prospected in recent years. Many of the buildings in the old town are well-preserved, and the town is now a well-known tourist attraction.

Geology. The central portion of the district is underlain by crystalline limestone of the Calaveras Formation (Carboniferous to Permian), which has many potholes and crevices that contained much rich gravel. The rest of the area is underlain by graphitic slate and schist. To the north and east several deposits of early Tertiary quartz-rich gravels were mined by hydraulicking and drifting. Those to the north are parts of the deposits of the ancestral Cosumnes River that extended west through the area and west-northwest towards the Fiddletown district. In places the gravels are capped by andesite. A few narrow goldquartz veins are found in the district.

West Point

See text under Calaveras County, p. 51.

Mines. Amador County: Amador-Columbus, Belden $400,000+, Black Prince $100,000+, Defender $100,000+, Elkhorn, Hageman, Jumbo, Lone Willow $100,000+, Newman $160,000+, Pine Grove, Pioneer Lucky Strike $300,000+, T.N.T.

White Oak Flat

White Oak Flat is in north-central Amador County about 10 miles northeast of Volcano. Several moderate-sized deposits of Tertiary channel gravels were mined years ago. There are several gold-quartz deposits including the Marklee mine, which was prospected recently. Bedrock consists of slate, chert, and quartzite. In places the gravels are overlain by andesite and rhyolite.

• Placer Mines of Amador County

• Allen Ranch. Henry and Weaver operated a dragline dredge on the Allen Ranch property on Sutter Creek Gulch during part of 1941. [1]

• American Flat Mine (placer). It is 3/4 of a mile southwest of Oleta, on American Flat channel. Course of channel is northeast and southwest. Workings consist of two incline shafts 300 feet deep, and a tunnel 900 feet in length. The width of channel from 100 to 200 feet wide, with a depth of gravel of 4 to 7 feet, lying on slate bedrock.

Arroyo Seco (Lambing) Mine (placer). It is 3 miles north of Ione. Idle.

Battle Mountain Mine (placer). It is 1 mile north of Volcano. Course of channel northeast and southwest, width of channel is 700 feet with a depth of gravel of 1 to 7 feet. Developments consist of tunnel 570 feet long. The bedrock is slate, and channel is capped with rhyolite-tuff 150 feet thick.

Clark Gravel Mine (placer). It is 2 miles south of Oleta. Course of channel is northeast and southwest. Capped with andesite flow 100 feet thick. Width of channel is from 60 to 300 feet. Depth of gravel from 4 to 7 feet and slightly cemented. Developments consist of two bedrock tunnels 800 feet in length.

Cleveland Consolidated Mine (placer). It is situated 1/4 mile northwest of Volcano. Course of channel is northwest and southeast. Capped with andesite flow 150 feet thick. Width of channel 60 to 150 feet, with a depth of gravel of 18 inches to 3 feet. The bedrock is slate. Developments consist of tunnel driven 417 feet west in slate bedrock, and then northwest on channel 1100 feet. The gravel is cemented, containing large boulders of quartz and quartzite. The gravel is first allowed to slack and then treated in a Dodge disintegrating mill.

Clinton Bar Mine (placer). It is 4-1/2 miles southeast of Pine Grove in canyon of Mokelumne River. Working the river bar, and drifting on old river channel.

Columbus Gravel Mine (placer). It is 4-1/2 miles northwest of Volcano. The channel has a northerly and southerly course and is capped by an andesite flow 100 feet thick. Width of channel is from 150 to 500 feet, depth of gravel being 2 to 8 feet. Gravel is partially cemented. Bedrock is a mica schist. Developments consist of two tunnels; north tunnel 400 feet long and south tunnel 200 feet in length.

Dry Creek Mine (placer). It is 5 miles north of Volcano; is worked by ground sluicing.

Elephant and China Hill Hydraulic Mines (hydraulic). This mine, near Volcano, was operated in 1932 by the K. D. Winship estate. One and one-half to 3 feet of gravel on decomposed slate bedrock are overlain by 40 to 45 feet of volcanic ash. Also controls the following claims: Elephant, China Hill, South Branch, Mallers, Hull and Ross, and Hodgdon Grant. Total of 645 acres. Water rights of Amador County to Tiger, Panther, Mill, and Antelope creeks; also owns Amador County Canal and McLaughlin ditches.

An extensive amount of hydraulic mining has been done on these placer holdings in past. In late years no mining operations have been carried on.

Fifty-Four Flat Mine (placer). It is 5 miles northwest of Volcano. Idle.

Forty-Nine Flat Mine (placer). It is 4 miles north of Volcano. Workings consist of two tunnels 1200 and 800 feet in length, with numerous lateral drifts. Channel capped with volcanic tuff 100 feet thick. Width of gravel is 100 feet with a depth of 2 to 6 feet. Bed rock is a soft mica schist. [12]

• Garibaldi Mine. Garibaldi Bros., of Volcano, operated a nonfloating wash plant at their mine on Pioneer Creek half a mile east of Volcano, intermittently during 1941. The yield from 33.200 cubic yards of gravel was 229 ounces of gold and 35 ounces of silver. [1]

• Gritton Mine (placer). It is 4 miles southeast of Volcano. In granodiorite formation. Shaft 125 feet deep. Tunnel 200 feet in length on the vein. Idle. [12]

• Horton Mine. This mine is in Jackson Valley, 5 miles south of Ione.

Indian Creek. Mountain Gold Dredging Company operated on Indian Creek 4 miles west of Plymouth with two dragline excavators.

Kent Property. During 1940, E. A. Kent of San Francisco operated two dragline dredges on Sutter Creek between the towns of Sutter Creek and Volcano.

Lancha Plana Gold Dredging Company. Operated on Jackson Creek near Buena Vista throughout 1940.

Lorentz Property. Lorentz operated on the Cosumnes River, 7 miles northwest of Plymouth, in 1941. [1]

• Pigeon Mine (placer). It is 1-1/2 miles northwest of Oleta. The course of channel is northeast and southwest, capped with andesite tuff 100 feet thick. Width of channel is 150 feet with depth of gravel from 4 to 8 feet. Development consists of bedrock tunnel 200 feet long. The gravel is cemented. Idle.

Posey Hill Mine (placer). It is 5 miles northeast of Volcano. Idle.

Rancheria Mine (placer). It is 5 miles northwest of Volcano. The channel is 200 feet wide with a depth of gravel of 6 to 8 feet, which is capped by an andesite tuff 150 feet thick. Bedrock tunnel 500 feet in length.

Red Cloud Mine (hydraulic). It is situated 12 miles southeast of Ione. Hydraulicking a 6-foot bank of loose gravel wash. [12]

• Rupley Ranch. This property is on Willow Creek 5 miles west of drytown. J. C. Pantle operated a dry-land dredge here from January 1 to October 15, 1942. He recovered 1290 ounces of gold and 183 ounces of silver from 230,000 cubic yards of gravel. [1]

Sunnyside Mine (drift). It is 2 miles south of Oleta. The gravel is partially cemented. Workings consist of a bedrock tunnel 700 feet in length. It is on the same channel as Clark mine. [12]

BUTTE COUNTY

Bangor-Wyandotte

Location. This district is in southeastern Butte County about 10 miles southeast of Oroville. It is an extensive area of placer deposits that occur in the vicinity of the old towns of Bangor and Wyandotte. The Honcut dredging district is just to the southwest.

History. The district was originally mined during the gold rush. Extensive drift and hydraulic mining was done from the middle 1850s through the 1890s. Bangor, founded in 1855 by the Lumbert brothers and named for their home town in Maine, was an important mining and staging center. Numerous Chinese were in the district from the 1870s through the 1890s. This and the adjoining Honcut district were dredged in the early 1900s and again in the 1930s.

Geology. Tertiary gravels covering a broad 3- by 8-mile area are believed to represent an ancient delta of the Tertiary Yuba River. The channel flowed northwest through Bangor and then west. Farther west are shore gravels of Pleistocene age.

The channel gravels were mined by drifting, the shore gravels by hydraulicking, and the Recent gravels by dredging. The channel gravels are as much as 150 feet thick, well rounded, well cemented, and consist of intrusive and metamorphic rock fragments with 10 to 20 percent quartz. The gold was flaky and often rusty. A number of coarse nuggets were found, one of which was reported to have weighed 14 pounds. Bedrock consists mainly of greenstone and amphibolite.

Drift mines. Bangor, Blue Lead, Catskill, Gray Lead, Turner.

Bidwell Bar

Location. Bidwell Bar is in southeastern Butte County about 10 miles northeast of Oroville and west of the junction of the South and Middle Forks of the Feather River. The district includes the Hurleton, Stringtown and Enterprise areas to the cast.

History. Gold was discovered here in 1848 by General John Bidwell, soon after Marshall's discovery at Coloma. News of this rich find spread, and there was a general rush to the Feather River region. Bidwell Bar, Long's Bar, Thompson Flat, Potter's Bar, Adamstown, and other settlements were soon established. All of these towns have long since disappeared, as the gravels were exhausted in a few years and the miners moved elsewhere. The old suspension bridge and a few remaining buildings later became a state park. Much of the area was inundated by Oroville Lake.

Geology. The district is underlain by amphibolite in the west and granite in the east. Most of the gold was obtained from Recent and Pleistocene gravels in and adjacent to the river. There are a few narrow gold-quartz veins.

Butte Creek

Location. Butte Creek is in north-central Butte County. It is chiefly a dredging district that extends along Butte Creek from about three miles southeast of Chico northeast to Centerville and Helltown, a distance of almost 12 miles. The Magalia district is contiguous on the northeast. The streams were placer-mined during the gold rush, and hydraulic mining and some drift mining of Tertiary gravels followed. The creek was worked with primitive power shovels and washing plants in the early 1900s. It was dredged from around 1902 to the early 1920s, again in the 1930s, and from 1945 to 1949.

Geology. The deposits consist mainly of stream and bench gravels in and along Butte Creek. They range from a few hundred feet wide at the upper end to nearly a mile at the lower end. The gravels are coarse, well rounded, and consist of andesite with some chert and minor quartz. Also there are Tertiary shore and bench gravels in the Centerville and Helltown areas. Dredging depths ranged from 13 to 35 feet, with an uneven bedrock. The last operations were reported to have yielded as much as 35 cents of gold per yard.

Dredging concerns. Butte Creek Cons., 1909-16, one bucket-line; Lancha Plana, 1941-49, one bucket-line; Pacific Gold Dredging Co., 1902-17?, two bucket-line; Piedmont Dredging Co., 1941, one Becker-Hopkins; Thurman and Wright, one dragline; Yuba Cons., 1941, one Becker-Hopkins.

- Butte Creek Consolidated Dredging Company.—This company began operations in the spring of 1909, and has an operating plant of one dredge. It is organized under the laws of the State of California, and is capitalized for $2,500,000. …

  The holdings of the company comprise an area of 1,450 acres, located in sections 24 and 25, township 22 north, range 2 east, and sections 19, 20, 17, and 8, township 22 north, range 3 east, on Butte Creek, near Diamondville, and about 10 miles northeast of Chico, Butte County, California. Out of the total area about 500 acres are considered dredgeable, but in order to acquire this the company was compelled to take over the non-dredgeable land as well. The ground was prospected by means of shafts and drills. The gravel lies on a sandstone bedrock, and ranges in depth from 13 to 28 feet. The entire deposit is difficult to handle with a dredge, owing to the many boulders and uneven bedrock. [2]

Cherokee

Location and history. This district is in central Butte County, 12 miles north of Oroville on the north side of Table Mountain and in the vicinity of the town of Cherokee or Cherokee Flat. It was so named for a party of Cherokee Indians who migrated here in the 1850s to mine gold. It has also been known as the Spring Valley district. Most of the output has come from the single large hydraulic mine, which is estimated to have yielded about $15 million. The town reached its heyday in the middle 1870s when it had a population of about 700.

Geology. The Tertiary placer deposits are associated with a west-trending channel. In this area the channel is in a trough about 700 feet wide. The sequence from bottom to top of the hydraulic pit is as follows: irregular greenstone gravel 5-10 feet thick that is lean in gold and contains local black clay streaks and minor basalt blocks; a rich 20- to 30-foot layer of coarse fresh blue gravel with large greenstone blocks, coarse and fine gold, small diamonds, and minor platinum (this layer yielded as much as several dollars per yard); several feet of decomposed gravel; 50 feet of sand and quartzitic gravel, the lower part of which yielded 25 cents per yard; 200 feet of clayey sand; and 50 to 75 feet of massive basalt.

Between 400 and 500 small diamonds were recovered from the gold-bearing gravels at the Cherokee mine. Several of the stones were more than two carats in weight and of good quality, but most were small and had a pale-yellow tinge. This is the best-known diamond-bearing locality in California.

Clipper Mills

Clipper Mills is in southeastern Butte and northeastern Yuba County about eight miles

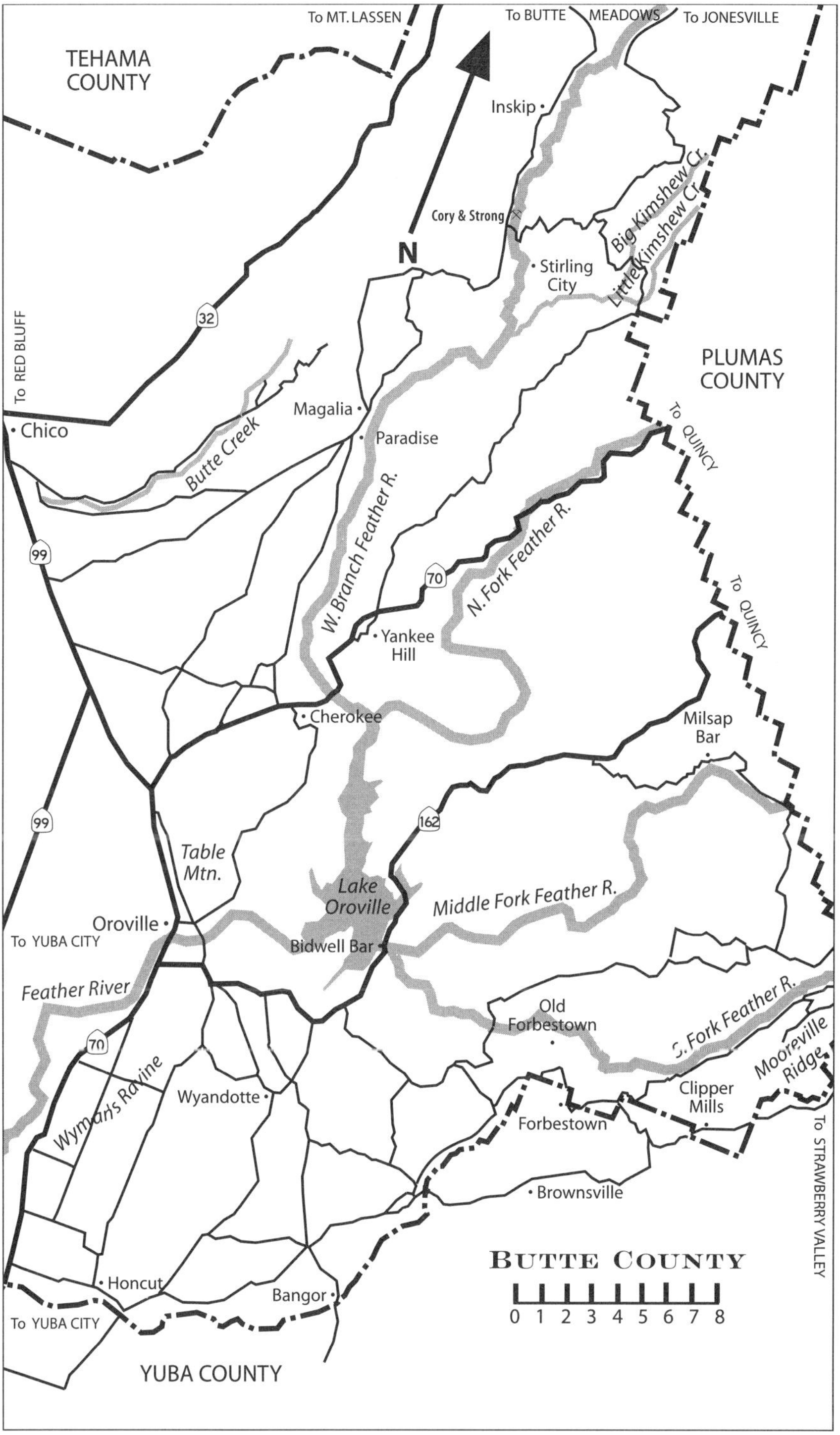
To MT. LASSEN
To BUTTE MEADOWS
To JONESVILLE
TEHAMA COUNTY
Inskip
Cory & Strong
N
Big Kimshew Cr.
Little Kimshew Cr.
Stirling City
32
To RED BLUFF
PLUMAS COUNTY
Magalia
Chico
Paradise
Butte Creek
To QUINCY
W. Branch Feather R.
N. Fork Feather R.
99
70
To QUINCY
Yankee Hill
Cherokee
Milsap Bar
99
162
Table Mtn.
Lake Oroville
Middle Fork Feather R.
Oroville
To YUBA CITY
Bidwell Bar
Feather River
S. Fork Feather R.
Old Forbestown
70
Mooreville Ridge
Wyman's Ravine
Wyandotte
Clipper Mills
Forbestown
To STRAWBERRY VALLEY
Brownsville
BUTTE COUNTY
0 1 2 3 4 5 6 7 8
Honcut
Bangor
To YUBA CITY
YUBA COUNTY

east-northeast of Forbestown. Several moderate-sized deposits of Tertiary channel gravels have been mined by drifting and hydraulicking. The Pratt and Gentle Anna drift mines and the Pittsburg Hill hydraulic mines have been the principal placer-gold sources. A few narrow goldquartz veins have yielded small but rich pockets. The district is underlain by slate, serpentine, and gabbrodiorite.

Forbestown

Location and History. The Forbestown district is in southeastern Butte County about 15 miles due east of Oroville. It includes the Feather Falls area. The Brownsville district is just to the south, and the Hurleton and Bidwell Bar districts are to the west. The area was placer-mined during the gold rush. At that time the South Fork of the Feather River, which drains the area, yielded huge amounts of gold. The town was named for Ben F. Forbes who established a store here in 1850. The Gold Bank mine was operated on a major scale from 1888 until 1904. Moderate mining activity continued in the district through the 1930s, and there has been intermittent prospecting, some by skin diving, since.

Geology and Ore Deposits. The area is underlain by massive greenstone, amphibolite schist and finegrained diorite. Granite lies to the north and gabbro-diorite to the south. There are a number of north-trending quartz veins and stringers. The ore contains free gold and considerable amounts of sulfides; tellurides have been reported. The milling ore is usually low in value (1/5 oz. or less per ton), but the ore shoots often had stoping lengths of several hundred feet. The sulfide concentrates were valued up to $100 per ton.

Mines. Carlisle, Denver, Gold Bank $2 million, Golden Eagle, Golden Queen $372,000+, Miller, Shakespeare $137,000+, Southern Cross, Williams.

Granite Basin

Location. The Granite Basin district straddles the Butte-Plumas County line about 30 miles northeast of Oroville and 30 miles west southwest of Quincy. It includes the Buckeye, Gold Lake, Milsap Bar, Soapstone Hill, and Merrimac areas. The area was placer mined during and after the gold rush. There was some lode mining here in the 1930s, and there has been minor prospecting since.

Geology. Several granitic stocks are intrusive into slate, quartzite and limestone of the Calaveras Formation (Carboniferous to Permian) and amphibolite. Exfoliation has formed several round granitic domes, of which Bald Rock is the most prominent. Serpentine, some Tertiary basalt and a few patches of Tertiary auriferous gravel are present. The quartz veins usually occur in the granite. The veins are narrow, but the ore bodies often are rich. Pyrite and galena are abundant.

Mines. Placer: Buckeye drift, Coquette, Horseshoe, Milsap Bar, Robinson.

Honcut

Location and History. This is a gold-dredging district in southwest Butte County along Honcut and Wilson Creeks northeast of the town of Honcut. 'The name comes from Hoankut, an Indian village once situated on the Yuba River just below the mouth of Honcut Creek. The Bangor district is just to the east. The creeks were first worked by hand methods in the early days. Bucket-line dredging began in 1909 and continued until around 1920. There was some dragline dredging in the district during the 1930s.

Geology. Pleistocene gravels and Recent creek gravels overlie bedrock of greenstone and green schist. Digging depths averaged about 20 feet. The dredged area covers about 1000 acres.

Honcut Creek District

- The Honcut Creek district is located on North Honcut Creek, in township 18 north, range 4 east, and about five miles northeast of Honcut, a station on the Southern Pacific Railroad. There are two dredges operating in the district, one belonging to the Kentucky Ranch Gold Dredging Company, a stock company, and the other to Laurence Gardella.

The Kentucky Ranch Gold Dredging Company began operations May 1, 1909, with one dredge. The company is incorporated under the laws of Arizona and is capitalized for $300,000. The holdings of this company comprise an area of about 1,000 acres, located in sections 25, 24, and 19, township 18 north, range 4 east, on North Honcut Creek, about 12 miles southeast of Oroville. The dredgeable area is estimated at about 200 acres. The gravel was prospected by means of drills, and is said to yield about 17 cents per cubic yard, the average depth to bedrock being about 18 feet; the bedrock is a schist.

The Gardella holdings are located on the Mahle property, almost joining the Kentucky Ranch Gold Dredging Company's holdings on the south. [2]

Inskip

Location. Inskip is in northeastern Butte County about seven miles north of Sterling City. The Kimshew district lies to the east and the Magalia district to the south. The area was active before and during the early 1900s.

Geology. There are a number of narrow gold-quartz veins in slate, amphibolite and greenstone. The ore bodies usually are small but contain abundant sulfides.

Mines. Bluebird, Cain, Excelsior, Fitzpatrick, Inskip, Lost Treasure, Midas, Rawhide, Walker, Wild Yankee.

Kimshew

Location. The Kimshew mining district is in northeastern Butte County and northwestern Plumas County about 10 miles northeast of Stirling City. It includes the Golden Summit area to the north.

Geology. There are a number of moderate-sized deposits of Tertiary and Pleistocene gravel that have been mined both by hydraulicking and drifting and a few narrow gold-quartz veins. Bedrock consists of slate and amphibolite to the west and a granitic stock to the east.

Mines. Brown, Carr, Cash Entry, Gallagher and Perkins lode, Golden Summit, Little Johnnie, Ream and Burnside, Reese and Jones, Snow, Wescott.

- *Cory and Strong Placer* is 2.8 miles by road north of Stirling City and contains 320 acres in the east half section 16, T. 24 N., R. 4 E., on the west side of West Branch of Feather River. A large lava-covered trough, locally known as the Mammoth Channel, crosses the property. Bedrock rims are 1500 to 2000 feet apart. Years ago, an adit was driven at a point 400 feet south of the north property-line to a length of 700 feet. It is in lava-wash supposed to be higher than the main channel. [1]

Magalia

Location. The Magalia district is in north-central Butte County 15 miles northeast of Chico. It is bounded on the west by Doe Mill Ridge and an the east by the West Branch of the Feather

River. It extends from Paradise on the south to several miles west of Powellton on the north. This district includes the placer deposits at Nimshaw, Forks of Butte, Mineral Slide and De Sabla and lode deposits at Toadtown. The Butte Creek dredging district adjoins this district at Centerville.

History. This region was extensively mined during and after the gold rush. The town was started in 1850 by E. B. Vinson and Charles Chamberlin. It was first known as Dogtown, renamed Magalia about 1862. The Magalia mine was discovered in 1855 and the Indian Springs mine in 1860. Large-scale mining continued until the 1890s; there was some activity from the early 1900s through the 1930s. There has been minor prospecting and development work since World War II. Some of the old mining properties have been made into housing subdivisions. The famous 54-lb. Willard, Dogtown, or Magalia nugget was found here in 1859. This is one of the more productive placer mining districts in the state. Several local residents have estimated the total output to be $40 million, but that figure is too high (author). Much of the output has come from drift mines.

Geology. There are a number of south-southwest-trending steep, narrow, and rich channels. The longest channel is the Magalia or Mammoth channel that flowed along the east side of the district. Other productive channels include the Dix, Emma, Little Magalia, Pershbaker, and Nugget channels. In the south portion of the district there are shore gravels. The gold was extremely coarse, and a number of other large nuggets besides the Willard were taken here. Bedrock is slate and greenstone with smaller amounts of serpentine. The channels are faulted in places with the downstream side being thrown up. Water has always been a problem in the drift mines. A few gold-quartz veins in greenstone are associated with diorite dikes.

Mines. Drift: Bader, Black Diamond, Cole, Cory, Dix, Emma $1 million+, Ethel, Genii, Indian Springs, Kelly Hill, Lucky John, Lucretia, Magalia $1 million, Mammoth, Mineral Slide, Nuggett, Oro Fino, Parry, Pershbaker, Pete Wood, Pitts, Princess, Royal, Steifer, Willard. Hydraulic: Centerville, Kohl, Red Hill.

Mooreville Ridge

Mooreville Ridge is in southwest Plumas and southeast Butte Counties. It includes the Camel Peak and American House areas. A number of Tertiary channel gravel deposits in the area have been mined largely by hydraulicking. The South Fork of the Feather River, which flows through the district, was mined also in the early days and skin divers are active in the area now. Bedrock in the district is granite, slate, serpentine, and amphibolite. The ridges are capped by Tertiary andesite and basalt. A few gold-quartz veins occur in the district.

Mines. Butte County: Dodson hydraulic, Golden Trant, Ludlam hydraulic, Walters. Plumas County: American House hydraulic, Browns Hill, Davis Point hydraulic, Fall River, Sanborn, Walters.

Morris Ravine

Location. This district is at the south side of Oroville Table Mountain three miles north of Oroville. It includes the Monte de Oro area. It is both a lode and placer-mining district. Morris Ravine and other nearby ravines were first placer-mined during the gold rush. Drift and lode mining began soon afterward and continued until around World War I. There was some work again during the 1930s. The Morris Ravine mine has been intermittently worked in recent years.

Geology. The bedrock in the district consists of amphibolite with smaller amounts of slate and phyllite. The slate contains fossil ferns. The bedrock is overlain by sedimentary and volcanic rocks of Oroville Table Mountain, a mesa-like hill consisting of thick beds of sands, tuffs, clays and auriferous channel gravels capped by black basalt. Fossil leaves also have been found in the clays.

Ore Deposits. The channel gravels are quartz-rich, well-cemented, and interbedded with sands and clays. The gold ranges from fine to coarse. A few small diamonds have been recovered here. The veins usually are narrow but the Banner vein has been mined to a depth of 1000 feet. The ore bodies are large but low in grade (1/10 ounce of gold/ton). Small rich pockets were mined also. Sulfides are spotty but often rich.

Mines. Placer: Monte de Oro, Morris Ravine, Perkins and Goodall, Yuba.

Oroville

Location. This district is in southwestern Butte County. It is mainly a dredging field that extends from just west of the city of Oroville southwest along the Feather River to a point about five miles due east of Biggs. The field is one to two miles wide and nine miles long.

History. Shallow placers were mined here during the gold rush. The area was settled in 1849: Oroville originally was known as Ophir City, but the name was changed in 1855. Around 1895, W. P. Hammon and others tested the area to determine the feasibility of mining on a large scale. They introduced bucket-line dredging in 1898, the first in California. The field was highly productive from 1903 to 1916; in. 1908 there were 35 dredges and 12 dredging companies active in the field. Output later declined, but dredging was done again from 1936 to 1942 and 1945 to 1952. The dredge field is now an important source of sand and gravel. The total output from dredging is estimated to be about 1,964,000 ounces of gold.

Geology. The gold occurs in river gravels and adjacent terrace gravels on the flood plain. The gravels rest on a bedrock of soft but compact andesite and rhyolite tuff. Coarse boulders, which become finer downstream, are present along with alternating sand layers. Digging depths ranged from 25 feet upstream to as much as 55 feet downstream. The gold was fine and occurred chiefly in the gravels. It was 915 to 930 in fineness. Dredge recoveries ranged from 15 cents to 25 cents per yard of gold at the new price. Minor amounts of platinum also were recovered.

Concerns. Butte Dredging Co., El Oro Dredging Co., Feather River Development Co., Gold Hill Dredging Co. 1938-50, Gold Run Dredging Co. 1906. Indiana Gold Dredging Co. 1908, Kentucky Ranch Gold Dredging Co. 1909, Natomas Cons. 1909-17, Oroville Dredging Ltd. 1906-16, Oroville Gold Dredging Co. 1941-44, Oroville Union Gold Dredge. Co. 1914, Oroville Water Light & Power Co. 1906-16, Pacific Gold Dredging Co. 1906-16?, Pennsylvania Dredging Co., Shasta-Butte Gold Dredg. Co. 1928-?, Viloro Syndicate Ltd. 1904-16?, Yuba Cons. 1935-52. Many other earlier concerns were consolidated around 1906.

• OROVILLE DREDGING GROUND

The Oroville dredging ground is considered to comprise about 7,000 acres of the present flood plain of Feather River, or a few feet above it, beginning a few hundred yards below town and continuing down for about 5 miles, where the ground gradually gets poorer, and thus more difficult for the dredgers to handle. The width varies from less than a mile to about 2 miles. The average depth of gravel is 25 or 30 feet, increasing to 40 feet lower down on the river. The gravel rests on a false bedrock consisting of a compact volcanic tuff containing no large lava pebbles or fragments. The sediments as a rule rise rather abruptly 50 to 70 feet above the flood plain, and consist of probably the same tuff covered with heavy Quaternary gravels on top, spreading in sloping plateaus. On slopes this gravel has in places been concentrated and shows evidence of old shallow diggings. Near Oroville and also lower down old Chinese workings are plentiful in the dredging ground, in part worked in open pits by Chinese pumps, in part by drifting. Much of the best ground is thus already exploited.

The deposits consist of heavy gravels, many cobbles being over 6 inches in diameter; even at a distance of 5 miles below Oroville, 6 to 8 inch cobbles are plentiful. At that place one diabase boulder 2 or 3 feet in diameter was noted. Gravel and sand alternate; sand, coarse gravel, or fine gravel may rest on the bedrock. Most of the gold is in coarse gravel; here and there some is contained in sand. The gold is fine, and some of it may be called flour gold. The tenor ranges from 12 to 20 cents to the cubic yard.

The dredging industry at Oroville has developed rapidly in recent years. In 1901 only 11 dredging machines were at work. In 1905 the product of 15 companies, with about 35 dredges, was $2,262,000. In 1909 13 companies with 26 dredges reported a product $2,743,000.

The bedrock at the Oroville dredging ground is a tuff that is pretty compact, though soft enough to be cut by dredge buckets. It contains small white pumice-like fragments, consisting of volcanic glass; in places, however, the fragments are very small and the tuff looks more like a compact clay. This tuff is exposed all along by dredging 20 to 40 feet below the surface; it also shows in the steep south bank above the tailings level 3 miles below Oroville, where it is capped by gravel, and on the road to Bidwell Bar 2 to 3 miles east of Oroville. On the Wyandotte road it outcrops in the low foothills below the gravel up to elevations of about 400 feet. It appears also one-fourth mile below Oroville bridge, on the north side of the river, below a 30 to 40 foot bench of Quaternary gravel. The contact of this Quaternary river gravel and the greenstone bedrock slopes gently westward, and the contact between greenstone and tuff seems to dip more steeply to the west.

This series of tuffs, sands, and clays extends south of Feather River for several miles; probably it underlies the Quaternary gravels of the whole foothill region between the Feather and the Yuba, and it is reported to underlie the gravels at the Yuba River dredging ground. Its thickness is unknown but probably increases rapidly westward. Below Oroville bore holes 80 feet deep have been sunk in it without finding any different material. Its correlation is somewhat difficult. It might be considered the exact equivalent of the Ione formation, but the fact that no tuffs have been found in the Ione formation at Table Mountain is against that correlation. The origin of its volcanic constituent is probably not the slope of the Sierra Nevada. It is more likely that ash showers from the neighboring volcano of the Marysville Buttes contributed to its formation and that it was deposited on the bedrock at the close of the Neocene, before the canyon of Feather River had been cut but after an active epoch of erosion of the older Ione formation. The bedrock relations at Oroville indicate pretty clearly that this series was deposited on the even Miocene slope corresponding to that of the Ione formation on the north side of the river, before the modern canyon had been excavated.

The Quaternary river gravels rest at Oroville on these beds in several fairly well defined benches. At Oroville there is a comparatively narrow, lowest bench 20 feet thick. The bedrock is 30 feet above the river, at an elevation of 160 feet above the sea. Above this there are two rolling benches which have an elevation of 340 to 430 feet, the latter being the maximum elevation of the Quaternary gravels on the Bangor road. [6]

• Dredging in the Oroville District.

Nearly all dredge mining in Butte County is carried on in the Oroville or Feather River district on land adjacent to the Feather River. The district, comprising a dredgeable area of about 6,450 acres, nearly two thirds of which has been already dredged, includes in its boundaries part of Oroville and extends six miles south of the town, ranging from one to several miles in width. Some minor dredging areas in Butte County are located near Oroville on Wyman's Ravine and Honcut Creek, tributaries of the Feather River, and on Butte Creek near Diamondville. As dredging in California began in the Oroville district, it may be of interest to give in detail some of the early history of the industry.

W. P. Hammon and Warren Treat were among the first to attempt working the gravels for gold on a large scale or by different methods than the sluice box or rocker. Treat, in the summer of 1895, sunk a pit about 100 feet square to bedrock, using a centrifugal pump to handle the water, and in spite of the heavy costs caused by the rehandling of the gravel several times by manual labor and for pumping, made a profit. Hammon ... had for some time thought of the then little known process of gold dredging in connection with same.

At the beginning of the year 1909 there were thirteen companies in the field, controlling about 6,450 acres and operating 30 dredges.

The Indiana Gold Dredging Company began operations in 1901 under the name of the Indiana Gold Dredging and Mining Company, and up to 1907 commissioned two dredges.

The holdings of the company comprise 163 acres located in sections 13 and 14, township 19 north, range 3 east, and sections 18 and 19, township 19 north, range 4 east, all on the east side of the Feather River and about one half mile to one mile distant from the town of Oroville. The dredgeable area comprises about 140 acres, of which more than 100 acres have been dredged. More than two thirds of the dredging area had been mined by Chinese and whites by the processes in use before the advent of the dredge; the rest is overflow land near the banks of the Feather River.

Butte Dredging Company.—Was organized under the laws of the State of Nevada. The company owns one dredge which was put in commission November 26, 1902, and dismantled in July, 1910.

The holdings of this company comprise an area of 80 acres located on the west side of the Feather River, in section 7, township 19 north, range 4 east, in Thermalite, and lots 134 to 137 of the townsite of Oroville. Most of the ground was mined by hand or hydraulic processes previous to dredging, and all of it is located on low, bar land, subject to overflow during flood times. The gravel, which is overlain by hydraulic tailing, was prospected by means of drills with test holes at intervals of about one to every four acres. It is a loose river gravel, carrying about 40 per cent of sand, and averaging 34 feet in depth to bedrock.

El Oro Dredging Company.—This company began operations in 1904 and up to 1910 has commissioned two dredges. The company was organized under the laws of Arizona.

The company holdings comprise an area of 218.54 acres, located on the east side of the Feather River about one mile inland, in sections 17 and 20, township 19 north, range 4 east. Practically all of this land was covered by placer pits as a result of the early Chinese mining previous to dredging. The gravel deposit is one of the most difficult to dredge in Oroville district; it is very tight and in places partly cemented, and carries throughout considerable clay. The ground was prospected by means of drills, with one test hole to every nine acres.

Ophir Gold Dredging Company.—This company has an operating plant of two dredges. It was organized under the laws of the State of Nevada. The holdings of this company comprise an area of about 211 acres, located in sections 17, 18, and 19 north, range 4 east, about one mile inland on the east side of the Feather River. Less than one half of this land was planted to orchard, etc., and most of it had been mined previous to dredging. The gravel was well prospected by drills, and averages about 27 feet in depth to bedrock; in character, the deposit is fairly compact, carrying in places considerable clay, but few large boulders.

Part of the holdings of this company formerly belonged to the Nevada Gold Dredging Company, which was originally the Central Gold Dredging Company.

Gardella Dredging.—This is a private concern owned by Laurence Gardella, of Oroville, California, and operating one dredge.

The holdings comprise an area of 40 acres, formerly planted to orchard, located in section 18, township 19 north, range 4 east, about one mile inland on the east side of the Feather River.

Gold Run Dredging Company.—Began operations in 1906 and has an operating plant of one dredge.

The holdings comprise an area of 122 acres, located in sections 20 and 29, township 19 north, range 4 east, on the east side of the Feather River, about one and one fourth miles inland. Practically all of this land was uncultivated and had been mined by hand previous to dredging.

The gravel was prospected by means of shafts, and like the El Oro deposit, it is very compact and in places partly cemented, carrying considerable clay and some large boulders. The average depth to bedrock is about 35 feet.

Viloro Syndicate, Limited.—This company began operations in 1904, and up to 1909, operated one dredge.

The holdings of the company comprise an area of 200 acres, located in sections 5, 19, and 30, township 19 north, range 4 east, on the east side of the Feather River, about one and one half miles inland. The dredgeable area comprises about 170 acres, of which about 150 acres had been previously mined by hand, and none of this ground was suitable for cultivation. The property was prospected by means of drills. The gravel deposit is compact, carrying medium coarse gravel, and in places, considerable clay. The average depth to bedrock is about 30 feet. A little platinum is recovered with the gold.

Oro Water, Light and Power Company.—This company began operations in 1903 and has an operating plant of five dredges.

The holdings of the company comprise an area of about 1,616 acres, of which about 1,300 are located on the east side of the Feather River in sections 18, 19, and 30, township 19 north, range 4 east, and in sections 25 and 33, township 19 north, range 3 east, township 18 north, range 3 east, and the rest on the west side of the river in section 24, township 19 north, range 3 east. The dredgeable area is estimated at about 1,400 acres, all of which is inland, and most was unsuitable for horticultural purposes.

The company took over the holdings of the Lave Bed Dredging Company and the Marigold Gold Dredging Company.

Pennsylvania Gold Dredging Company.—This company began operations in 1902, and has an operating plant of one dredge.

The company holdings comprise an area of 155.5 acres, located on the east side of the

Feather River in section 25, township 19 north, range 3 east, and formerly the property of the California Dredging Company, which had an operating plant consisting of one dipper dredge. Most of the ground had been mined by hand previous to dredging and none of it was cultivated with success. The property was prospected by means of drills. The gravel is medium coarse and free from clay, averaging about 28 feet in depth to bedrock.

Oroville Dredging, Limited.—This company has an operating plant of six dredges and one machine shop.

In the Oroville district the holdings of the company comprise an area of about 1,346 acres, located in sections 18 and 19, township 19 north, range 4 east, on the east side of the Feather River, extending as far as one mile inland, and in sections 23, 24, 25, and 26, township 19 north, range 3 east, on the west side of the Feather River, extending as far as two miles inland. With the exception of about 200 acres, none of this land has been under cultivation, and practically all of it was mined previous to dredging. The properties were carefully prospected by means of drills. The gravel averages about 12-1/2 cents per cubic yard, and in character ranges from a clean, loose, river wash, to a heavy bench gravel, in places partly cemented, and at times carrying considerable clay. The average depth to bedrock is about 30 feet.

In the Bear River district, the company owns 993 acres, located in the Bear River Basin in Placer and Yuba counties.

The Boston and California Dredging Company began operations in 1902. It took over the holdings of the Leggett and Wilcox Company, which had an operating plant consisting of two dredges. Part of the holdings, the Wilcox tract, was land owned by the company, the Leggett tract being land leased by the company on a royalty basis. Both of these companies were promoted by W. P. Hammon, and operated under his direction, the personnel consisting mainly of Boston men.

The Boston and Oroville Mining Company began operations in 1901. This company was among the first started by W. P. Hammon. It had at one time, an operating plant of four dredges, but at the present time has only one dredge in active operation.

Oroville Dredging and Exploration Company began operations in 1902. The holdings are located on the west side of the Feather River.

The Feather River division of the Natomas Consolidated of California has an operating plant consisting of two dredges. On January 1, 1909, the Natomas Consolidated of California took over the holdings of the Cherokee Gold Dredging Company, having an area of 200 acres, and an operating plant of one dredge; and the Feather River Exploration Company, having an area of about 900 acres and five dredges, some of which were abandoned.

The total holdings of the company comprise an area of about 1,364 acres, located in sections 24, 25, and 26, township 19 north, range 3 east, on the west side of the Feather River, extending as far as one mile inland; and in section 3, township 18 north, range 3 east, and section 24, township 19 north, range 3 east, on the east side of the river, extending as far as one and one half miles inland. With the exception of about 80 acres, none of this land has been under cultivation, and most of it was mined by hand previous to dredging. On some of this land the first dredging operations in the Oroville district were carried on. The gravel averages from 9 cents to 14 cents per cubic yard, the upper part of the property being the richer, and in character the gravel is a clean river wash, in places overlain by several feet of heavy loam. The average depth to bedrock ranges from 25 to 40 feet.

The Feather River Exploration Company was the first company to begin dredging operations in the Oroville district. It was originally started by W. P. Hammon and Captain Thomas Couch, and to these enterprising men much credit is due for the rapid development of gold dredging in California since 1897. The holdings of this company were taken over by the Feather River Development Company in January, 1906.

Pacific Gold Dredging Company.—This company began operations May 1, 1906, and has an operating plant of four dredges. It is a subsidiary company of the Yukon Gold Company. The holdings comprise an area of 750 acres, located in sections 2, 3, and 4, township 18 north, range 3 east, and sections 33 and 34, township 19 north, range 3 east, on the east side of the Feather River, extending as far as 1-1/2 miles inland. With the exception of about 100 acres, none of this land has been under cultivation, and a great deal of it was mined by hand previous to dredging. The gravel was well prospected by means of drills, and averages in depth to bedrock about 28 feet, and yields an average of about 9 cents per cubic yard; in character it is mostly a clean, loose gravel, overlain by fine loam and sand to a depth of several feet.

This company took over the holdings of the American Gold Dredging Company, having an operating plant of two dredges, and an acreage of 275 acres; and the Kia Oro Gold Dredging Company with an acreage of 153 acres and one abandoned dredge.

Leggett Gold Dredging Company.—This company began operations in March, 1904, and had an operating plant of one dredge; it was a close corporation, with James H. Leggett, president.

The holdings comprised an area of about 75 acres located in section 18, township 19 north, range 4 east, on the east side of the Feather River, about one half mile inland. The land was part of the Leggett ranch, all planted to orchard previous to dredging.

Wyman's Ravine District

The Wyman's Ravine district is located in township 19 north, range 4 east, about 4 miles southeast from Oroville. The dredging lands are located along Wyman's Ravine, which empties into Prairie Slough, a tributary to South Honcut Creek, which flows into the Feather River.

There are three companies in the district, the Leggett Gold Mining Company, one dredge; L. & J. Gardella, one dredge; and the Garden Ranch Gold Dredging Company, with one dredge, now abandoned.

The Leggett Gold Mining Company began operations in August, 1909. The holdings of this company comprise an area of about 520 acres, located north of Palermo, on Wyman's Ravine, in section 5, township 18 north, range 4 east, and sections 32 and 33, township 19 north, range 4 east. Of the total area about 350 acres are estimated as dredgeable. Of this area about one quarter was previously mined by hand and none of it has been under cultivation. The gravel was prospected by means of drills and averages about 10 feet in depth to bedrock. In character the gravel is a fairly clean, loose wash, carrying few large boulders and very little clay, and in places is overlain by a sandy loam. The gravel is said to yield 15 cents per cubic yard.

The Gardella property, located in Wyman's Ravine district, is owned by L. & J. Gardella, father and son partnership. The holdings of this company comprise an area of about 440 acres, located in sections 32, 28, 27, township 19 north, range 4 east, on the Fahey and Humphrey property, about five miles southeast of Oroville. The total dredgeable area is estimated at 175 acres, of which about 20 acres have been dredged. The gravel was prospected by means of drills, and averages about 10 feet in depth to bedrock, and in character is fairly loose, carrying, in places, large boulders and some streaks of clay, and is overlain by a sandy loam. It is said to average 16 cents per cubic yard.

The Garden Ranch Dredging Company has an area of about 160 acres, located in sections 22 and 23, township 19 north, range 4 east, on the upper end of Wyman's Ravine, and about four miles southeast from Oroville.

The Garden Ranch dredge has been idle for some time, having worked out the property. [2]

Yankee Hill

Location. This district is in east-central Butte County about 15 miles northeast of Oroville. It is fairly extensive and includes the Concow and Big Bend areas.

History. The streams and surface placers were first worked during the gold rush. For a time the locality was known as Rich Gulch and Spanishtown. In those days much gold was recovered from the North Fork of the Feather River, and a diversion tunnel was driven through Big Bend. Numerous Chinese miners reworked the old placer tailings later on. Lode mining began in the 1850s, and there was much activity during the 1890s and early 1900s. The Surcease mine was worked on a major scale from 1933 to 1942, and copper was mined at the nearby Big Bend mine during World War II. The estimated output of the district is slightly more than 100,000 ounces of gold.

Geology. A northwest-trending belt of slate and quartzite four to five miles wide, with some limestone that is part of the Calaveras Formation (Carboniferous to Permian), crops out in the central part of the district. Interbeds of amphibolite and serpentine lie to the north. Granodiorite stocks are to the east and southeast.

Ore Deposits. A number of quartz veins contain some free gold and often abundant sulfides, especially chalcopyrite. The veins are in the metamorphic rocks. Milling ore commonly averages 1/2 ounce of gold per ton, much of the values being in the sulfides. The surcease vein has been mined to a depth of more than 1000 feet. A gold-bearing barite vein occurs at the Pinkston mine.

Mines. Berry Creek, Bunker Hill, Evening Star, Hearst, Madre de Oro, Pinkston, Porter, Rainbow, Sunbeam, Surcease $1 million+, Treasure Hill.

CALAVERAS COUNTY

• Placer Gold Production. The total amount or value of placer gold produced in Calaveras County since 1848 is not known. Records kept prior to publication of "Mineral Resources of the United States" by the U.S. Geological Survey in 1880 were incomplete and often inaccurate, and from 1880 to 1896, published statistics did not separate placer and lode gold production. Since 1896, placer operations have accounted for approximately one-fifth of the total gold production from Calaveras County; for the preceding 48 years, however, the proportion must have been higher. Julihn and Horton (1938, p. 21) have estimated that the value of placer gold produced in Calaveras County prior to 1896 was "probably as much as $75,000,000". Production from this date through 1936 amounted to approximately $9,100,000 (Julihn and Horton, 1938, p. 15); placer production in the 20 years following possibly totaled about $7,600,000.

History. The first significant recovery of gold in Calaveras County occurred in 1848, after gold had been discovered in the Mokelumne and Stanislaus Rivers and their tributaries. It was in the summer of this year that a party of miners, including James Carson, George Angel, and John and Dan Murphy, entered the county from the north. The group prospected the streams as they worked southward, and finally disbanded near what is now Angels Creek. Carson went south to the present site of Melones, while Angel and others remained and began panning the surface gravels. The town of Angels Camp grew up, and by 1853 had a population of about 4500. Shortly afterward local placer mining died out and was replaced by lode mining.

When the group of miners disbanded, the Murphy brothers went to the present site of Vallecito, which was called "Murphy's Old Diggins". Shortly after, they moved to "Murphy's New Diggins" (Murphys), and the Old Diggins came to be known as Vallecito, a name used by the local Mexican population. The ground at Murphys was so rich that local miners' regulations in 1849 limited each man to a single square claim 8 feet on a side. Mining was done by panning until the Fall of 1849, when rockers were introduced from Tuolumne County. In January 1853, the Union Water Company completed a 15-mile ditch that brought water to Murphys from the Stanislaus River. In September 1859, a bedrock flume to Murphys Flat was completed; this facilitated drainage and removal of tailings, and allowed the placer mines to be worked to a depth of 35 feet. Placer mining in the area flourished until 1863.

San Andreas was sparsely settled by a few Mexicans in 1848, although the first mining was not done until the Spring of 1850, by some Americans. Water was brought in by the Union Company in 1852, and the mines were in full operation a year later. In 1855, a Mr. Murray struck buried gravels of a Tertiary river channel on the north point of Douglas Hill, and sank a shaft. Although the last good surface placers were worked in 1857, there were more than 80 mines working the channels within 1 mile of town in July 1859.

Mokelumne Hill was settled in 1848. Three years later, a company of Frenchmen discovered a rich deposit on what later was called French Hill. This area, as well as those surrounding, was mined by hundreds of "coyote holes"—irregular, sparingly timbered holes that were dug to a depth of 100 feet, and around which excavated dirt and gravel were piled.

Discovery of placer gold was responsible for the founding and rapid settlement of many other towns in the county, including Poverty Bar, Middle Bar, and Big Bar on the Mokelumne River; Campo Seco, which was rescued from abandonment by the discovery of copper after the surface placers had been worked out; and Camanche, Jenny Lind, Milton, Calaveritas, Dogtown, West Point, and Railroad Flat.

The earliest placer mining in Calaveras County was by means of the miner's pan. In the Fall of 1849, rockers were introduced at Murphys, and soon were in use throughout the county. This device was quickly supplemented and replaced by the long tom, introduced from Georgia, and the sluice box. Hydraulic mining was developed at Nevada City, and the use of this high-volume, low-cost method of mining soon spread throughout the Mother Lode region. By 1854, Calaveras County had 17 ditch systems, totaling 325 miles in length, to supply water for the hydraulic mines. Drift mining began in the county at San Andreas, in 1855, and increased steadily in importance until around the turn of the century.

Hydraulic mining remained the dominant method of gold recovery until shortly after 1884. This was the year that Judge Sawyer of the United States Circuit Court handed down a decree restraining the North Bloomfield Mining Company from discharging debris into the South Fork of the Yuba River. Hydraulic mines were steadily closed by injunctions based on this decision until, in 1893, the federal Caminetti Act created the California Debris Commission and forbade hydraulic mining that would injure navigability or lands in territory drained by the Sacramento and San Joaquin river systems. Some hydraulic mines in Calaveras County continued operating for a few years after this law was passed, but gradually their debris dams were filled, and by 1905 this form of mining had virtually ceased.

The first successful dredging operation in California was started in 1898 at Oroville by an open-link bucket dredge. Connected-bucket dredges were introduced in 1901. By the end of the next decade, dredging had spread to and was firmly established in Calaveras County. Operations centered around the Camanche and Jenny Lind areas. Dragline dredging came into prominence in Calaveras County and the State as a whole in the early 1930s. The value of gold recovered by this method reached its peak in 1941. Although it never threatened to displace bucket-line dredging as a source of income, it did allow the working of many small properties that were not rich or large enough to warrant the capital investment of a bucket-line dredge. The U.S. Bureau of Mines reports that in Calaveras County during 1941, there were eleven dragline dredges, one connected-bucket dredge, one drift mine, and eight miscellaneous placer operations that included the use of power shovels, bulldozers, and stationary washing plants. War Production Board Order L-208, issued in October 1942, curtailed placer mining activity during 1943-45. The period 1946-54 in this county saw the use of seventeen dredges, one stationary washing plant, one suction dredge, and five drift mines.

Mining methods. Early placer mining in Calaveras County involved the use of simple hand equipment: a pick, shovel, and gold pan or batea. However, the capacity of a gold pan in the hands of a skilled operator is between only one-half and one cubic yard of gravel in 10 hours. The rocker and long tom, therefore, were natural steps in the effort to increase the amount of gravel a man could wash in a day. Next to be used was sluicing, a method in which gravel is washed by passing it through a long, three-sided, wooden trough called a "sluice."

Sluices are built with a slight gradient, and contain riffles along the bottom behind which particles of gold are trapped. In ground sluicing, gravel is excavated and carried through sluices by water that is not under pressure. Mining is done by allowing a stream of water to fall over a bank, washing material into the sluice, or by periodically releasing water from a storage reservoir situated upstream from the gravel. Monitors may be used to supplement the action of ground-sluice water, but if they perform most of the work, the method becomes hydraulic mining.

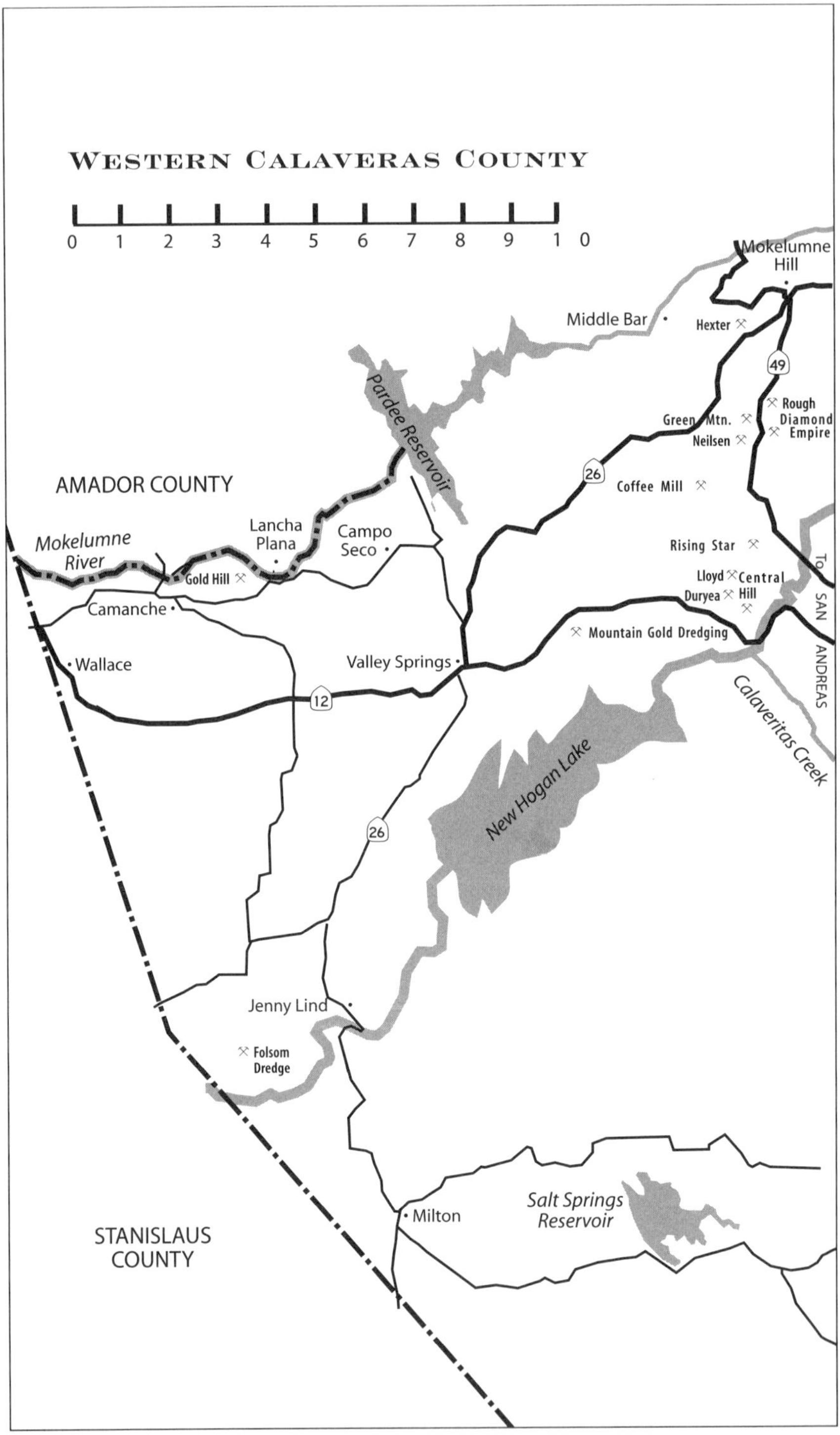
WESTERN CALAVERAS COUNTY
0 1 2 3 4 5 6 7 8 9 10
Mokelumne Hill
Middle Bar
Hexter
49
Rough Diamond
Empire
Green Mtn.
Neilsen
Pardee Reservoir
26
Coffee Mill
AMADOR COUNTY
Lancha Plana
Campo Seco
Mokelumne River
Gold Hill
Rising Star
Lloyd
Central Hill
Duryea
To SAN ANDREAS
Camanche
Mountain Gold Dredging
Wallace
Valley Springs
12
Calaveritas Creek
New Hogan Lake
26
Jenny Lind
Folsom Dredge
Milton
Salt Springs Reservoir
STANISLAUS COUNTY

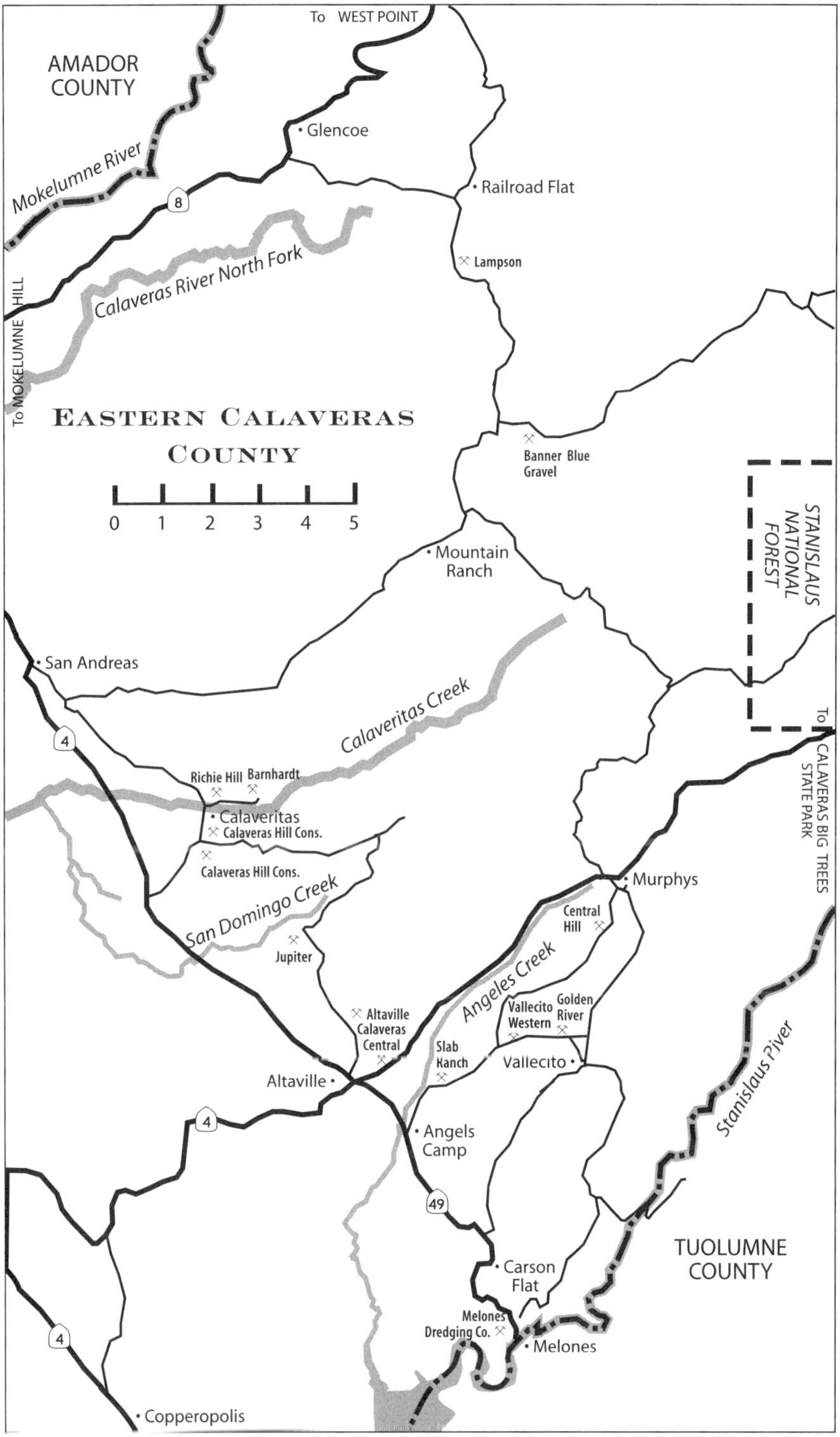

To WEST POINT
AMADOR COUNTY
Glencoe
Mokelumne River
Railroad Flat
8
Calaveras River North Fork
Lampson
To MOKELUMNE HILL
EASTERN CALAVERAS COUNTY
0 1 2 3 4 5
Banner Blue Gravel
STANISLAUS NATIONAL FOREST
Mountain Ranch
San Andreas
Calaveritas Creek
4
To CALAVERAS BIG TREES STATE PARK
Richie Hill
Barnhardt
Calaveritas
Calaveras Hill Cons.
Calaveras Hill Cons.
Murphys
San Domingo Creek
Central Hill
Jupiter
Angeles Creek
Vallecito Western
Golden River
Altaville Calaveras Central
Slab Ranch
Vallecito
Altaville
4
Angels Camp
Stanislaus River
49
TUOLUMNE COUNTY
Carson Flat
Melones Dredging Co.
Melones
4
Copperopolis

© Gem Guides

In hydraulic mining, a jet of water under high natural pressure is directed through a nozzle called a "monitor" against a bank of gravel in order to disintegrate it and wash it through sluices. Almost all types of exposed placer deposits can be worked by hydraulicking if sufficient water is available, but the capacity of the operation becomes limited if the gravel is cemented or clayey, if the bedrock is flat, or if disposal of tailings and water by gravity is very difficult. Hydraulicking is a low-cost method of mining: it yields a larger production per man than any other method except dredging, and requires a smaller initial investment than dredging. Considerable development work often is necessary in preparing a deposit for hydraulicking: water reservoirs, several miles of flumes, hundreds of feet of iron pipe, one or more monitors, heavy sluice boxes, and a tailing dam were standard equipment in most Sierra hydraulic mines.

A deposit of gravel usually is opened at its lower end, to facilitate movement of gravel and water by gravity. After the deposit is opened, the gravel bank is undercut by the monitor, allowing the overlying material to cave into the pit. As the cutting and sweeping capacity of a monitor generally exceeds the carrying capacity of the water used, extra water is provided by a flowing stream, called "by-wash," which is directed through the pit and into the sluices. Large boulders are either disintegrated by blasting, removed by a derrick, stacked on the pit floor by hand, or swept through the sluices. It was common practice to use more than one monitor in a pit. Usually one or more high-pressure monitors were used to cut the bank, while a lower-pressure monitor was used simultaneously to sweep gravel toward the sluices.

Deposits of gold-bearing Tertiary stream gravel in the Sierra Nevada, capped by volcanic rock or other barren overburden, have been exploited by means of drift mining. A body of such buried gravel may be opened by a shaft or an adit. Drifts usually are run on bedrock upstream from the adit portal or the shaft station so that water readily drains from the mine. Common practice is to prospect the buried channel by a drift along the lowest part of its trough; crosscuts are run to either rim, and raises occasionally are put up to explore for gold-bearing strata above bedrock. The actual plan of mining depends chiefly upon the width of gravel to be mined and the configuration of the bedrock.

In a narrow channel, the drift and faces may be simultaneously advanced across the full width of the pay streak. In a wide channel, one or two parallel drifts may be run to the limit of the deposit, and then small blocks of gravel are mined, retreating toward the shaft. A modified room-and-pillar method sometimes is used, but more commonly the roof is supported by widely spaced stulls. If the gravel is loose, extensive timbering may be necessary. In mining, "breasts" or long faces of gravel are either picked by hand or drilled and blasted, then shoveled into ore cars and trammed to the portal or shaft station. This type of mining is called "breasting."

Most Sierran Tertiary gravels are partially or wholly cemented, and it is necessary to disintegrate the gravel before washing it to recover the gold. This commonly was done by stamp mills before the early 1900s, but since then it has been accomplished at most mines by passing the gravel through a scrubber-trommel. A trommel consist of a slightly horizontally inclined revolving steel cylinder on the inside of which iron rails or other lifting devices have been attached. The gravel is introduced through the elevated end of the cylinder and is lifted and dropped several times before it passes onto the screen or punched-plate section of the cylinder at the lower end. Broken gravel that has passed the trommel is washed in combinations of sluices, screens, jigs, and tables. If the gravel is tightly cemented, it may be re-cycled through a crusher-and-screen unit before washing.

Extensive flat-lying, shallow, river-bar and alluvial plain gravel deposits in which relatively small amounts of gold are evenly distributed are especially suited to dredging. Enough water must be available to float the dredge and wash the gravel; the dredged ground must be right enough to hold water in the dredge pond; and the bedrock should he relatively smooth and decomposed.

Other conditions favorable for dredging are the absence of overburden, cemented gravel, clay beds, large boulders, and living or buried timber. A temperate climate, allowing year-around activity, and availability of power and transportation facilities also are important.

Connected-bucket dredges were used almost exclusively until the early 1930s, when drag-line dredges were introduced. A connected-bucket dredge can dig up to three times as deep as a dragline and is able to handle harder bedrock and tighter gravel. On the other hand, a dragline dredge requires less capital, is more flexible in its operation, is easy to move after a deposit has been worked, and requires only a relatively small pond to float the washing plant.

The most common type of connected-bucket dredge used in California has a washing plant consisting of a trommel, sluices, tables and/or jigs, and a stacker to dispose of oversize from the trommel. Digging is accomplished by a bucket ladder mounted in the bow of the dredge, which is swung slowly from side to side by winches and cables mounted near the stern and connected to deadmen on shore. An overhead winch above the ladder enables it to be raised and lowered as conditions require. Because the bucket ladder is in the bow and the tailing stacker in the stern, the dredge continually moves forward, carrying its pond with it. Sand is discharged into the pond through sluices, but does not impair the motion of the dredge.

Dragline dredges consist of a dragline on the shore that excavates gravel and dumps it into a washing plant that floats on a small dredge pond. The dragline retreats as it cuts away the bank on which it stands, and the washing plant floats forward to stay within range of the dragline. The washing plant has a flowsheet similar to that employed on a connected-bucket dredge.

Other related methods of working placer deposits include suction dredges and stationary washing plants supplied by power shovels and trucks. Suction dredges have been used in California, but never successfully. Principal disadvantages are their low capacity and the ease with which feed lines become obstructed by clay, boulders, or sunken wood. Power shovels, trucks, and stationary or movable washing plants may be used if there is insufficient water for hydraulicking, if the deposit is too small to warrant the investment necessary for a hydraulic mine or dredge, or if the topography and condition of the bedrock do not permit other types of mining. A movable washing plant eliminates trucking charges and permits easy disposal of tailings. A stationary plant is structurally more sound and better designed than a movable plant, and in addition, feed to the gold-recovery equipment does not fluctuate when the shovel moves or stops, because the stationary plant has a larger feed storage bin.

Geology. Gold-bearing placer deposits that are economically important in Calaveras County are of three types: residual soil and weathered rock, Quaternary stream gravels and transported soils, and Tertiary stream and shoreline gravels.

Auriferous residual soil and weathered rock were commonly first worked by small-scale placer mining methods, and sometimes by hydraulicking. Such deposits were shallow, however, and soon depleted. If the bedrock from which they were derived was sufficiently rich, lode mining was started. Many lode mines in Calaveras County began in this manner.

Quaternary streams and soils in meadows and "flats" were another easily accessible source of gold in the early days of the Gold Rush. Railroad Flat and Murphys Flat are Quaternary deposits that were worked by hand methods and ground sluicing in the 1850s and 1860s. Gravels in Recent streams were mined on a small scale at Carson Creek, Middle Bar, and Dogtown, and on a larger scale later at Jenny Lind and Camanche.

Because of the diminishing yield from Quaternary surface placers since the early 1900s, Tertiary channels and shoreline gravels have assumed relatively greater importance in the economy of gold mining in the county. The Tertiary channels of Calaveras and other Sierran counties emptied into the Ione Sea, the shoreline of which has not been located precisely, although gravels along a northwest-trending belt between Wallace and Jenny Lind probably represent delta deposits of the Tertiary river.

The Tertiary Calaveras channel traces its origin to Cretaceous time, when the drainage was controlled by a series of subdued, parallel, north-northwest-trending ridges, breached in a few places to allow the streams to reach a shallow sea lying to the west. This system persisted into the early Tertiary, and was thus inherited by the Eocene streams. The subtropical climate and relative stability of orogenic forces during the early Eocene permitted deep weathering and decomposition of the rocks. Thus, when the Tertiary Sierran uplift began, the rejuvenated streams became loaded with fine quartz sand and pebbles which originated from the erosion of lenses and beds of chert and veins of gold-bearing quartz, together with clay derived from decomposed feldspar in the deeply weathered rocks. Finer particles in the stream load were washed into the sea, and today are found in the Ione formation as large deposits of clay interbedded with quartz sand.

Explosive volcanic activity beginning at least as far back as Oligocene time resulted in rhyolitic ash-falls over much of the northern Sierra Nevada. Rivers were dammed, forming numerous small lakes in which streams deposited weathered and altered ash which now is seen as layered "pipe-clay" overlying auriferous stream gravel Newly developed rivers flowed over a pervasive volcanic cover, and were repeatedly interrupted by ejections of rhyolitic and, in Mio-Pliocene time, andesitic lava and mudflows. Because of this, the buried Tertiary channel systems are quite complex, generally consisting of a series of individual channels that are superimposed on and crosscut one another. The age of any such series of channels may range from late Cretaceous (?) to Pliocene. Prospecting for and mining buried placer gold deposits in this environment must be done with care, as not all the channels are gold-bearing.

Fragments of rhyolite and andesite were caught up in new and old streams, and were carried toward the ocean along with fragments of quartz, granite, and metamorphic rock. Consequently these stream-channel deposits can be classified on the basis of their constituent rock particles as prevolcanic, intervolcanic, and postvolcanic; they can be further classified with respect to the periods of rhyolitic and andesitic volcanism, which were distinct from each other and were probably separated by a period of quiescence. This widespread volcanic activity did not cease until late Pliocene or early Pleistocene time. The latite of Table Mountain, the source of which probably was in Tuolumne County, is the most recent product of this activity preserved in Calaveras County. It overlies much of the Cataract channel, which, because of its westward trend, probably is younger in age than the Calaveras system of channels.

A time of extensive earth movements followed during the Pleistocene, as the Sierran block was once more uplifted and tilted westward. Within the western foothills, smaller faults displaced some of the buried Tertiary channels and changed their apparent grade. Accelerated stream flow resulting from the westward tilting of the Sierran block eventually established a drainage system that cut westward across the general structural trend; some of these new streams cut deep and rugged canyons that are little changed today. Buried Tertiary stream gravel was exposed and partially removed by these new rivers, and gold contained in it was pirated and washed into the newer streams. Many of the very fine particles of gold were swept to the edge of the Great Valley and deposited with mixed Quaternary gravel to form the present dredging grounds.

The Tertiary channels of Calaveras County have been discussed in detail by Storms (1894), Lindgren (1911), and Julihn and Horton (1938), and in less detail by others.

W. H. Storms published the only detailed map of the Tertiary channels in Calaveras County that attempted to cover the whole county. Lindgren's maps were confined to the Mokelumne Hill-San Andreas and Dogtown-Murphys areas; Julihn and Horton published a detailed map of the channels between Altaville and Vallecito. Lorin Clark (1954) and J. H. Eric et al. (1955) published detailed geologic maps of the Calaveritas and Angels Camp quadrangles, in which the disposition of erosion remnants of Tertiary river gravel were accurately noted. None of these authors has been in complete agreement with the others. Small-scale maps of the major channels have been published by C. S. Haley (1923), Don Steffa (1932), and Olaf Jenkins (1932). [5]

Altaville

• Jupiter (San Domingo, Keefer) mine

Location: SE1/4 sec. 7, and NW1/4 sec. 17, T. 3 N., R. 13 E., M.D.M., 3 miles north of Altaville.

Ownership: Walter V and Norma M. Valente, Angels Camp, California.

This property, located on the Central Hill channel, was worked with a 100-foot bank as the Keefer hydraulic mine in 1880 and as the Jupiter mine in 1882. Some drifting was later done here prior to 1896. The mine operated as the Jupiter and later the San Domingo from 1898 to 1902(?), when more than 226,500 cubic yards of gravel were hydraulicked. It was again active as a drift mine in 1903, when an 1100-foot adit was completed. In 1908 approximately 69,000 yards of gravel were hydraulicked. The property remained inactive until a 600-foot adit was driven upstream N. 70° E. during 1934-36.

The channel here trends almost due west; it is steep sided and 125 feet wide. Slightly cemented gravel containing some boulders rests on a slate bedrock. About 2 feet of bed-rock were mined with the gravel during the drifting operations of 1934-36. The washing plant at that time consisted of a 4 1/2 - by 18-foot trommel-screen, a sluice box, and a stacker. [5]

Alto

This district is in the southwest corner of Calaveras County about six miles south of Copperopolis. It was named by the author for the Alto mine, a major source of gold in the area. The region was first placer-mined during the gold rush, when the gravels underlying nearby Tuolumne Table Mountain were worked by drifting. The Alto mine was discovered in 1886 and operated on a large scale until 1907. It has an estimated total output of $1 million. The district is underlain by slate of the Mariposa Formation (Upper Jurassic) with some interbeds of massive greenstone. The gold occurs in thin quartz veins or with disseminated pyrite in the greenstone. Placer gold was recovered from Eocene quartzitic gravels overlying the slate. Some of the gravels are capped by latite porphyry of Tuolumne Table Mountain.

Angels Camp

Location. This district is in southwest Calaveras County in the vicinity of the town of Angels Camp and Altaville. It is an important part of the Mother Lode belt and lies between the San Andreas district to the northwest and the Carson Hill district to the southeast. It is both a lode and placer district, but the lode mines have been more productive.

History. The streams in the area were mined shortly after the beginning of the gold rush. The town was founded in 1848 and named for Henry Angel, who had established a trading post here. Rich surface ores were mined in the oxidized zones in the 1850s, and most of the important veins were discovered at that time. By 1885 Angels Camp had become one of the major gold-mining districts in the state. The Utica Mining Company was organized in the middle 1850s, and for the next 40 years the Utica mine was a major source of gold. From 1893 to 1895 this mine yielded more than $4 million worth of gold. All of the major mines were shut down during World War I. There was some activity in the district again during the 1930s, and the Calaveras Central and Altaville drift mines have been intermittently prospected during the past 15 years. This district has an estimated total output of at least $30 million, and it may be considerably more.

The colorful jumping frog jubilee held each year at nearby Frogtown, the county fairgrounds, is based on Mark Twain's tale, *The Jumping Frog of Calaveras County.* He is supposed to have first related the story at the Angels Hotel.

Geology. This district is underlain by a series of northwest-striking beds of amphibolite and chlorite schist, phyllite, greenstone, and metagabbro. To the east and west are slate, impure quartzite, and micaceous schist. In the north and northeast part of the district is the west- and northwest-trending Tertiary Central channel, which is a tributary of the ancestral Calaveras River. The channel gravels are cemented, contain abundant quartz, and are overlain by rhyolite tuff and andesite.

Ore deposits. The lode deposits consist of massive quartz veins, zones of parallel quartz stringers and bodies of mineralized schist and greenstone. The ore contains disseminated free gold and auriferous pyrite. Usually the gold is in fine particles, although occasional high-grade pockets containing coarse gold have been found. Calcite, talc, ankerite, and sericite, are commonly present in the ore. The milling-grade ore usually averaged 1/7 to 1/5 ounce of gold per ton, but the ore bodies were scores of feet in thickness, hundreds of feet in length, and were mined to depths of several thousand feet.

The deposits occur either in the amphibolite and chlorite schist or phyllite. There are three principal vein systems. In the system on the west the veins are in phyllite. In the center one the veins are along the west margin of a northwest-trending belt of metagabbro. In the eastern system, which contains the famous Utica mine, the ore deposits are in amphibolite and greenstone.

Mines. Drift: Altaville, Calaveras Central 22,000 ounces+.

- Altaville mine

  Location: NW1/4 sec. 21, T. 3 N., R. 13 E., M.D.M., 1 mile north of Angels Camp. Ownership: Alta Mining Company, Inc., Altaville, California; George Schmauder of Altaville, lessee. The Altaville drift mine was first developed during 1939-42, when a 358-foot vertical shaft was sunk, from the bottom of which about 400 feet of drifts were driven. Between 1942 and 1949 the mine was inactive. From 1949 until the spring of 1953, the Alta Mining Company and later the Gold Bar Mining Company operated the mine intermittently. In 1950 and 1951, 300 feet of drifts were driven from the old workings, and 1200 feet were driven on an upper level at 341 feet. George Schmauder has leased the property since early 1953. In April 1956 pumping ceased, and the mine was allowed to fill with water.

This mine is on the Central Hill channel, about a mile downstream (northwest) from the Calaveras Central drift mine. The gravel is well cemented and consists of boulders and cobbles of chert, quartz, granitic rock, and schist; it rests on greenstone and schist of the Calaveras formation. The bedrock is about 350 feet below the surface; the total width of the channel is unknown. Gold is 888 to 900 fine, and ranges in size from flour to flat, pea-sized nuggets. Small amounts of platinum also have been recovered.

Calaveras Central mine

Location: Portions of secs. 21, 22, 23, 26, 27, and 28 (480 acres); main working shaft (Reiner) is in SW1/4 NE1/4 sec. 28, T. 3 N., R. 13 E., M.D.M., about 1 mile north of Angels Camp.

Ownership: Calaveras Central Gold Mining Company, Ltd., c/o Harry Sears, Trustee, Angels Camp, California.

The Calaveras Central mine is a consolidation of the Victor, McElroy, Aetna, and Slab Ranch properties, which were acquired by the Calaveras Central Gold Mining Company, Ltd., in the years 1931 to 1933. Each of these properties was prospected and developed before the turn of the century, although only the McElroy was worked intensively from the first (see Victor and Slab Ranch).

The Victor mine was purchased in 1926 from the Victor Canal and Mineral Co. by the Calaveras Central Mining Corporation, which carried on some early development work. In 1930, Calaveras Central Gold Mining Co., Ltd., was incorporated in Nevada, and the following year leased a few hundred acres, including the Victor or Reiner mine, from the Calaveras Central Mining Corporation. Other properties were acquired shortly thereafter, until the company controlled mining rights to the McElroy, Aetna (Peirano), Slab Ranch (Calmo), and E. W. Johnson Ranch properties. The 300-foot Victor (Reiner) shaft was reconditioned for use as the main working shaft, and the 240-foot Aetna shaft was retimbered and connected underground with the main workings to provide ventilation and an emergency exit.

Geological features of the Victor, McElroy, Aetna, and Slab Ranch mines are closely related. Gravels include those in at least three pre-volcanic channels, several poorly defined intervolcanic channels, and Recent surface gravel. About 500 feet east of the Slab Ranch mine shaft, the bedrock rifts of the Tertiary Central Hill river channel are less than 500 feet apart. The channel widens rapidly downstream, however, and within a mile forms a shallow basin about 1-1/2 miles wide. The Tertiary river flowed west at the Slab Ranch mine, but on entering the basin it swerved to the northwest, and by so doing cut several distinct channels separated at places by bedrock ridges. Three bedrock channels have been explored, and a fourth may lie to the northeast. The Aetna is the youngest of the known channels cut by the Tertiary Central Hill River, and at one point the Aetna and the main channel have apparently merged. The Central Hill channel also cuts the No. 5 channel in two places. The fourth channel to the northeast might represent an older bed of the Tertiary Central Hill River.

The Aetna channel, which is thought to have entered the basin from the south through a gap in the rimrock, might not be one of the main channels of the Tertiary river. Its gravels contain a larger proportion of quartz sand and boulders than do the other channels, and its gold is finer-grained and less worn; many pieces with quartz matrices indicate travel over a short distance. In contrast, gold in the Central Hill channel is well worn, is of large size, and generally is identical in character with that from the Vallecito-Western and Golden River mines upstream. The Central Hill channel thus appears to have been formed by the Tertiary Calaveras River (Julihn and Horton, 1938, p. 45).

Several intervolcanic channels, of which little is known, overlie the bedrock channels in the basin. The McElroy is the only such channel to have been worked. It enters the basin from the south and west of the Aetna channel, and may have derived its gold from erosion of auriferous quartz veins in the Angels Camp-Altaville area. No lava flows are found in the basin, but two distinct layers of rhyolite tuff and a later one of andesitic tuff are recognized. Gold generally is found within the lower 5 feet of gravel and the upper 2 feet of bedrock, which is slate and schist of the Calaveras formation. The width of pay gravel ranges from 50 to 70 feet in the No. 5 channel, and from 150 to 250 feet in the Central Hill and Aetna channels. The gravel of the Central Hill channel consists of tightly cemented, well-rounded pebbles, cobbles, and boulders of granite, granodiorite, and quartz. The gold is between 885 and 900 fine because of its silver content, and much of it is larger than 10 mesh in size. Most of the gold is well worn. At places it is mixed with gold of less fineness that possibly was introduced by other channels. In both channels, it is associated with considerable pyrite and a little black sand. Gold in the No. 5 channel is smaller in size than in the other two channels, and its vertical distribution through the gravel is greater. There is little bedrock enrichment, but layers of sand as much as 30 feet above bedrock are gold-bearing.

Starting in June 1931, a crosscut from the station of the main shaft was driven 520 feet northeast, where No. 5 channel was intersected. This channel, which carried fine and medium-coarse gold, was explored for 1600 feet upstream, where it is partially cut off by a deeper channel, apparently part of the Central Hill channel. From 1934 to 1936, inclusive, most activity was confined to the Central Hill channel, which was followed downstream 2700 feet from the main shaft. The Aetna channel cuts through this channel 1500 feet from the shaft. These three channels are the principal ones that have been developed; approximately 30,000 feet of workings have opened more than 4000 feet of channel. Development work continued intermittently until the mine was closed in 1942. Gravel from the old tailings was processed and used for plant-mixed road surfacing in the early 1940s. Rehabilitation of the property began in 1947. Since 1950, a 600-tons-per-day mill, new surface equipment, and a 110-foot steel headframe have been installed, some of the old workings have been reopened, and the main and safety-exit shafts have been retimbered. The total yield of the mines in which the Calaveras Central Gold Mining Co., Ltd., has an interest is reported to have been 2000 ounces of gold from 1900 to June 1931, and 20,000 ounces in the 6 years following. Since 1946, development work has yielded only a small amount of gold. From June 1931 to 1937, about 185,000 tons of material were mined, of which 25 percent was waste bedrock. The average recovery during this time was $4.94 per ton. Moreover, re-treatment of 12,000 tons of fine tailing yielded an additional $1.35 per ton (Julihn and Horton, 1938, pp. 43, 47-48).

Most of the underground drifts have a cross-section of 6 by 7 feet, and require little or no timbering. Mining equipment includes lightweight jack-leg drills, mucking machines, drag scrapers, two 1/2-ton ore cars, and a compressed-air locomotive. A chisel bit for drilling and 40-percent powder for blasting are used to advance the drifts. The mine makes about 250,000 gallons of water per day.

The mill is designed to separate a clean gravel for road-building material as a byproduct of the gold recovery. Auriferous gravel is hoisted to the surface through the main shaft and tripped into a 250-ton circular ore bin on the headframe. The gravel is fed to a 2-inch grizzly; the oversize is crushed to minus-one inch and screened to 1/4 inch. Screen oversize goes to a belt conveyor and is stacked for future use as clean gravel; the undersize is passed over wooden riffles and sent to a Yuba jig. Minus-2-inch gravel from the grizzly is fed to a 5- by 8-foot inclined scrubber-trommel. Slotted discharge openings in the wall allow passage of minus-1/8 inch material; oversize is sent to the belt conveyor and stacked with gravel from the crusher. Trommel undersize is passed over wooden riffles and the Yuba jig. A concentrating table makes a final cleaning of the sand, which is then stockpiled for future use.

As work at the mine since 1950 has been limited to underground rehabilitation and surface improvements, the mill has been operated on only a pilot-plant scale.

Slab Ranch (Calmo) mine

Location: NW1/4 and SE1/4 sec. 27, T. 3 N., R. 13 E., M.D.M., 1-1/2 miles northeast of Angels Camp. Ownership: c/o P. N. Alexander, 1567 Hiawatha Avenue, Stockton, California.

This property, situated on the Central Hill channel, was prospected in the 1860s, but was not seriously worked until 1922. By late 1923 a vertical shaft 45 feet deep had been sunk. The Calmo Mining and Milling Company, later known as the Mound City Mining Company, sank two vertical shafts 167 and 115 feet, from which more than 8000 feet of drifts and crosscuts are said to have been driven in gravel and in bedrock in search of quartz stringers. This period of intermittent activity lasted from 1924 to 1933. The first shaft to be sunk was the Calmo shaft, downstream and north of the presently existing Slab Ranch shaft. Approximately 1500 feet of drifts and crosscuts were driven, principally in the bedrock rims, and gold valued at about $25,000 was produced from the Calmo. After that the mill and mining equipment were moved to the Slab Ranch site, and a shaft was sunk in a narrow portion of a canyon in the Central Hill channel. No production from this shaft has been recorded. In 1933, this property was acquired by the Calaveras Central Gold Mining Co., Ltd.

The channel trough at the Slab Ranch site is between 70 and 125 feet wide, and at the Calmo site it is somewhat wider; gravel on the dump contains cobbles of schist, quartzite, basalt(?), and rhyolite. A small dump at the caved Calmo shaft, which was sunk on the north rim of the channel, is composed chiefly as subangular cobbles of schist. The Slab Ranch washing plant consisted of a 3- by 25-foot scrubber-trommel and sluices.

Victor (Reiner, Reiner, etc.) mine

Location: NE1/4 sec. 28, T. 3 N., R. 13 E., M.D.M., one-half mile north of Angels Camp. Ownership: Calaveras Central Cold Mining Company, c/o Harry Sears, Trustee, Angels Camp, California.

This property was first located in 1895, and reportedly was sold for $50,000 that same year. The shaft, which now is the main shaft of the Calaveras Central drift mine, was sunk in 1904-05. The mine was worked intermittently by the Reiner Mining Company from 1908 until 1912, when fire destroyed the shaft and a 20-stamp mill. Work was resumed, and by mid-1913 bedrock was reached. About 1000 feet of drifts were driven by 1915, but a heavy flow of water in the lower levels closed the mine. The Victor Mining Company dewatered it and did some drifting in 1920-21. A small amount of platinum was recovered during this time from auriferous blue clay at the bottom of the 320-foot shaft. The Victor Land and Mineral Company worked the mine intermittently during 1922-26; drifts were driven east to the Central Hill channel, and small blocks of gravel averaging $5 to $6 per ton were mined south and northwest of the shaft. The property was purchased by the Calaveras Central Mining Corporation in March 1927, and remained idle until it was acquired by the Calaveras Central Gold Mining Company, Ltd., in mid-1931. [5]

Calaveritas

Location. This is a placer-mining district in central Calaveras County, in the vicinity of the old town of Calaveritas. It includes the Old Gulch area to the north. At one time there were many Mexican miners in the area.

Geology. The district is underlain by slate, graphitic schist, quartzite, limestone, green schist and granodiorite. Numerous gravel patches, remnants of the west-trending Fort Mountain channel of Tertiary age, were mined by hydraulicking and sluicing during the early days. The recent stream gravels were worked by drag-line dredges during the 1930s. There are a few narrow gold-quartz veins.

Mines. Barnhardt [Sec. 27, T 4 N., R. 13 E.], Calaveritas Hill Consolidated [Sec. 35, T 4 N., R. 12 E.], Oro Fino [Sec. 30, R. 4 N., T 13 E.], Railroad Hill [Sec. 2, R. 3 N., R. 12 E.], Richie Hill [Sec. 26, T. 4 N., R. 12 E.].

Camanche-Lancha Plana

Location. This district extends along the Mokelumne River from the vicinity of Lancha Plana, Amador County, west through the Camanche-Wallace area, Calaveras County, to Clements, San Joaquin County, a distance of about 12 miles.

History. The creeks were first worked during the early part of the gold rush, and hydraulic mining of the terrace gravels followed. The town of Camanche was named for a town in Iowa. Later, the Chinese mined the river and reworked the old tailings. From 1904 to 1923, the river was dredged on a large scale by the American Dredging Company. Dragline dredging was done during the 1930s and bucket-line dredging from then until 1951. Much of the region was recently inundated by the Camanche reservoir. The output from dredging is estimated to be about $10 million.

Geology. The deposits consist of unconsolidated gravels in and adjacent to the Mokelumne River and floodplain deposits. Tightly packed shore gravels of Eocene age are found near Wallace. In the west portion of the dredging field, the gravels ranged from six to 50 feet deep while, in the east, they were six to 25 feet deep. Bedrock is clay and volcanic rock. Values recovered by dredging ranged from 10 to 25 cents per yard. The gold was fine grained and 850 to 900 in fineness.

Operations. American Dredging Co., 1904-23, three bucket-lines; Camanche Placers Ltd., 1935-, one bucket-line; Gold Gravel Products, 1935; Gold Hill Dredging Co., 1936-51, two bucket-lines; Lancha Plana Gold Dredging Co., 1926-40, one? bucket-line; Wallace Dredging Co., 1935-40, two bucket-lines.

• Atlas Gold Dredging Corporation

Location: Secs. 11, 12, 13, and 14, T. 4 N., R. 9 E., M.D.M., 1 mile southwest of Camanche.

Ownership: Company inactive since January 5, 1949.

A dragline and floating washing plant worked on 240 acres leased from Calaveras Gold Dredging Company between late 1935 and early 1937. The dragline had a 2-1/2-yard bucket; the washing plant consisted of a 33-1/2-foot trommel, 700 square feet of sluices, Bendelari jigs, and a 70-foot stacker.

An average of 20 feet of Ione shoreline gravel covers about half of the property. Sampling was said to have revealed 4 million cubic yards of gravel averaging 15 cents per yard in fine gold, much of which was rusty (Julihn, 1938, p. 88). A small proportion of platinum also was found in the black-sand concentrates. The gravel is tight but infrequently cemented, and contains no large boulders. It rests on a bedrock of gray volcanic tuff. [5]

Campo Seco-Valley Springs

Campo Seco and Valley Springs are in northwestern Calaveras County. Substantial amounts of gold have been produced in this district from a variety of mineral deposits. These deposits include Recent stream gravels, quartz-rich gravels of Eocene age, narrow quartz veins, and massive copper and zinc sulfide deposits of the noted Penn mine, from which gold was recovered as a by-product.

The older quartzitic gravels were mined by hydraulicking and ground sluicing in the early days and later by drift mining. The Recent gravels were worked by dragline dredges. The Penn mine is credited with a production of 60,000 ounces of gold.

The principal rocks underlying the district are green schist, slate, serpentine and greenstone. The quartz-rich gravels which overlie these rocks occur in patches and are the remnants of an Eocene river channel system. Some of these gravels are capped by andesite. The massive sulfide deposits at the Penn mine and the gold-quartz veins are in greenstone and schist.

Hodson

Location. This district is in the Sierra Nevada west gold belt in southwestern Calaveras County a few miles west of Copperopolis. A belt of lode-gold deposits extends from the site of the old town of Hodson northwest through Salt Springs Valley, a distance of about 10 miles. At one time this district was also known as the Felix district.

History. Small-scale placer mining probably was done here during the gold rush, and the mining of rich surface pockets soon followed. The district was highly productive during the 1890s and early 1900s when the Royal mine and other properties were worked on a large scale. The 120-stamp mill at the Royal mine, erected in 1903, was one of the largest mines in California. The mines were active again during the 1930s and early 1940s. Copper ore from Copperopolis was treated at the Mountain King Mill during World War II. More recent exploration work, including diamond drilling, has been done in the district, but very little of it has been gold mining.

Geology. On the west side of the mineralized belt are northwest-trending beds of slate of the Mariposa Formation (Upper Jurassic). Metavolcanic rocks, chiefly massive greenstones and amphibolite of the Logtown Ridge Formation (Upper Jurassic), are on the east. The central portion has been intruded by numerous serpentinized bodies in or adjacent to the northwest-trending Hodson fault zone.

Ore Deposits. The deposits consist of large lowgrade bodies of mineralized schist and greenstone known as "gray ore," which contain some disseminated free gold, auriferous pyrite and minor amounts of other sulfides. The deposits are associated with the Hodson fault. One of the larger gray ore bodies is several thousand feet long, 500 feet wide, and has been mined to an inclined depth of several thousand feet. Some high-grade pockets have been taken from quartz veins and stringers containing free gold and sulfides.

Mines. Butcher Shop, Empire, Gold Knoll, Gold Metal, Mountain King $1 million, Pine Log, Ranch, Royal $5 million+, Wilbur Womble.

Jenny Lind

Location. This district is between the towns of Jenny Lind and Milton in western Calaveras County. It extends west into eastern San Joaquin and northeastern Stanislaus Counties. The area was first worked during the gold rush, and later hydraulicked. There was dredging here from 1903 until about 1940 and also small-scale lode mining. The district has yielded more than 100,000 ounces of gold.

Geology. The gold values are in river gravels and floodplain deposits in and adjacent to the Calaveras River. There are older terrace and shore gravels, some of which are overlain by hardpan. In places hydraulic mine tailings overlie the gravels. Dredging depths ranged from 20 to 40 feet, with the average nearer 20 feet. Recovered gold values ranged from 10¢ to 30¢ per yard and hydraulic tailings were around 10¢ per yard. There are a number of narrow gold-quartz veins in greenstone in the eastern portion of the district.

Dredging Operations. Butte Dredging Co., Calaveras Gold Dredging Co. 1903-16, El Oro Dredging Co., Isabel Dredging Co. 1908-25?, Milton Gold Dredging Co. 1935-?

Mokelumne Hill

Location. Mokelumne Hill is in northwestern Calaveras County. It is both a placer- and lode-mining district and includes the Chili Gulch, Old Woman's Gulch, and Golden Gate Hill areas.

History. The streams in the area were placer-mined early in the gold rush. Mokelumne Hill, first known as Big Bar, developed as a mining camp in 1848. Chili Gulch was first known as Chilean Gulch, from the large number of Chilean miners who worked here. They were discriminated against and often were forced to leave the mining regions. There was also a number of French miners in the district who had disputes with the Americans. Later in the 1870s Chinese miners were active in this district in great numbers.

Large amounts of gold were recovered by hydraulicking and drifting, but output declined in the 1870s. The Quaker City, Boston, and other lode mines yielded substantial amounts of gold from the 1880s until about 1900. Mining was done in the district again in the 1930s, and there has been intermittent prospecting since. The gravels at Chili Gulch are now mined for aggregate. Many of the buildings in the old town of Mokelumne Hill are well preserved.

Geology and Ore Deposits. A complex system of Tertiary channels, extending south and southwest from Mokelumne Hill, included eight distinct channels and remnants of several others, which range from Eocene to Pliocene in age. The so-called Chili Gulch, Stockton Hill, and Deep Blue channels have been the most productive. These contain a high percentage of quartz pebbles and boulders. In places they also contain large clear quartz crystals, some of which are piezoelectric grade. Usually the channels are not more than a few hundred feet wide. Bedrock consists of slate, greenstone, and graphite schist.

The gold-quartz veins occur in the slate and greenstone and are up to 50 feet thick. The gold occurs in the native state and is associated with small amounts of pyrite. Several dacite volcanic domes crop out in the district, and to the northeast a granodiorite stock is exposed.

Mines. Placer: America, Chappellet, Concentrator, Coffee Mill, Duryea, French Hill, Gopher, Green Mountain, Happy Valley, Hexter, Mosher, Neilsen, North Star, South Diamond, What Cheer, Werle.

• Church Union Mine (Kraemer) in sec. 26, T. 5 N., R. 11 E., M.D.M., is owned by J. J. McSorley and Thos. E McSorley, Mokelumne Hill. It includes 80 acres of mineral rights and 30 acres surface rights on the same ground, containing the Coffee-Mill channel, which runs in an east and west direction. According to J. J. McSorley, 20 acres of this ground is virgin, and should yield about an ounxe of gold per cubic yard for a 6-foot cut on bedrock. A tunnel 1000 feet long has been driven in bedrock beneath the channel. [1]

• Coffee Mill (Gleason, Golden Gate) mine

Location: SW 1/4 sec. 26 and NW 1/4 sec. 35, T. 5 N., R. 11 E., M.D.M., on cast slope Golden Gate Hill, 4 miles southwest of Mokelumne Hill. Ownership: Eunice N. Van Winkle and Edmund J. Stocker, 2631 Robbindale, Stockton, California.

The Coffee Mill drift mine, on the Kraemer channel, was first active in 1866, when a 400-foot inclined adit was driven. A pay streak 100 feet wide was struck, and by 1868 a 10-stamp mill had been erected. Production of gravel, in which coarse nuggets were found, reached 200 carloads per day but the mine was closed by excessive water in 1869 and again in 1871. A new inclined adit and main drift were driven 1500 feet during 1880-82. The mine was closed briefly, and in 1884 yet another adit was driven a distance of 800 feet. The property then was idle until 1892, when a 200-foot vertical shaft and a 25-foot east-trending drift tapped 6 feet of blue cemented gravel. A steam hoist was used at this shaft in 1896. The Coffee Mill was shut down before 1900, and was not actively developed thereafter except for retimbering of a 900-foot tunnel in 1915.

Duryea (Shaw) mine

Location: NE 1/4, sec. 1 3, T. 5 N., R. 11 E., M.D.M., in Chili Gulch, 1 1/2, miles south of Mokelumne Hill. Ownership: Not determined.

This property was first active as the Shaw hydraulic mine from 1866 to 1871. James Duryea worked it from 1872 to 1876, using a monitor with a 3-inch nozzle to wash a bank 160 feet high, which was regularly loosened by blasting. Overburden became increasingly difficult to handle, so a 5-stamp mill was purchased in mid-1876, and drifting was begun. This drift mine remained active until the mill was sold in 1911. It has not been developed since.

The Duryea channel courses through this property and trends southwest. It is 100 feet wide, and contains 30 to 40 feet of rounded quartz gravel and clean sand overlain by 10 to 20 feet of compact quartz sand, rhyolitic flows and ruff, and andesitic ruff and breccia. This mine is said to have yielded $500,000 by hydraulicking.

Underground workings include a 2000-foot and a 700-foot adit, a 300-foot crosscut, several shorter adits, and breasts as much as 14 feet high.

Green Mountain (Calaveras Crystal, McSorley Crystal) mine

Location: S1/2NE1/4, and N1/2SE1/4, sec. 24, T. 5 N., R. 11 E., M.D.M., about 2 miles south of Mokelumne Hill, north of and adjacent to the Rough Diamond drift mine. Ownership: R. P. M. Davis, 666 Rudd Street, Vista, California.

The Green Mountain mine was first active around 1874 when it was worked as a drift mine. By 1884 two adits, one 1200 feet long and another to the north 800 feet long, had been driven, although work had been confined to the northern adit since 1877. In 1885 the northern part of the property was worked by two east-trending adits 300 feet apart and each 900 feet long, connected by ventilation drifts. Near the east rim of the channel drifts were run 60 to 70 feet north and south. Gravel between drifts was breasted to a height of 6 feet at the farthest point, from which the work was carried toward the adits. The cemented auriferous gravel was treated in an eight-stamp mill at the rate of 70 carloads per day.

Large quartz crystals and opalized wood were reportedly found prior to 1896, and 12 tons of crystal valued at $18,000 were mined and shipped in 1897-98. Between 1895 and 1900, a different portion of the property was intermittently hydraulicked by two operators, who removed a total of about 11,500 yards of gravel. In 1908, the mine failed to yield gold, but $10,000 worth of gem quality quartz crystal was recovered. The Green Mountain mine was idle between 1910 and World War I, but by the end of that war had produced 22 tons of crystals, the largest of which was 6 feet long and weighed 2800 pounds. The next recorded activity was in 1925, when a 700-foot drift was driven on the Green Mountain Extension; a small amount of gold was produced in 1926. In 1934 intermittent exploration for quartz crystals ceased and was not resumed until May 1943, when the property was operated as the Calaveras Crystal mine. By June of that year the exploratory drift was 675 feet long, and 300 pounds of quartz crystal had yielded 90 pounds of saleable material. Quartz production continued through 1944, but none has been reported since. In 1945 a small amount of gold was recovered.

The Green Mountain and Rough Diamond properties are on the Tunnel Ridge channel, which flowed southwest at the time its gravel was deposited. This channel is truncated 900 feet north of the mines by a younger channel that contains a thin section of basal gravel capped successively by rhyolitic ash and andesitic cobbles and ash. The Tunnel Ridge channel cuts greenstone and slate of the Calaveras group and is filled entirely by gravel, which here is 1200 feet wide and 285 feet thick. The lower 125 feet of gravel has a matrix of sand and clay and is poorly cemented; the upper portion has a clean sandy matrix and is well cemented. In the lowest 10 feet of gravel, the average size of cobbles is 5 to 6 inches, although boulders as much as 3 feet in diameter are common on the bedrock surface. The gravel itself is in individual beds a few inches to 6 feet thick, and is composed chiefly of well-rounded quartzite, vein quartz, and quartz schist, with lesser amounts of altered slate, greenstone, mica schist, and serpentine.

In the vicinity of the mines the main channel axis describes an arc; north of the mines it trends west, and to the south it trends south. A small tributary stream joined the main channel from the northwest immediately southwest of the southernmost workings; another small tributary may have joined the channel from the east just north of the Rough Diamond mine. A feature of economic importance in the mines is a small cliff located from 125 to 200 feet cast of the channel axis. Its height ranges from 5 to 15 feet, and its slope ranges from nearly vertical to 45° W. Only a few quartz crystals have been found east of this cliff, and these were close to the ravine that entered the main channel from the cast and may have entered the channel through this ravine. No crystals have been found west of the channel axis. The width of the crystal-bearing gravel is 200 feet in the Rough Diamond mine and only 70 feet in the Green Mountain mine. Crystals have been found as much as 20 feet above bedrock, although they are most abundant in the lower 6 feet of gravel. The majority of crystals recovered have been in gravel rather than sand, but are not confined to any particular layer.

The distribution of the crystals, together with their almost complete lack of wear, suggests that their source is a nearby hydrothermal vein that was cut and eroded by Tertiary streams. There is no evidence to indicate whether any part of this vein still exists, or whether it might still contain quartz crystals.

The workings of the Green Mountain mine in 1944 consisted of about 1000 feet of southeast- and south-trending drifts and crosscuts. Many hundreds of feet of older workings, of which there is little or no record, have been caved for several decades.

Hexter mine

Location: NW 1/4 sec. 13, T. 5 N., R. 11 E., M.D.M., 1 mile southwest of Mokelumne Hill. Ownership: John B. Fershtand, 1501 Summit, Fort Worth 3, Texas.

The main haulageway of this drift mine was driven S. 30° E. for 1400 feet during 1896 to 1899, but it was worked only intermittently during the following decade. The adit was retimbered in 1912 and reached a length of 3100 feet in 1914. That same year a 200 tons-per-day mill was erected and a 130-foot ventilating shaft was sunk. The adit was 3800 feet long when the mine closed in 1917. Other workings include drifts 500 feet north and 500 feet south along the channel.

Pay gravel of the Duryea channel, on which the Hexter is located, is 7 feet thick and 125 feet wide. It is slightly cemented and capped by flows of rhyolite and andesite. Soft opals of sub-gem quality have been recovered from this mine. In 1914 gravel was treated in a rotary mill, concentrating table, and 70 feet of sluices.

Neilsen placer

Location: NW1/4, sec. 25, T. 5 N., R. 11 E., M.D.M., about 2 1/2 miles south of Mokelumne Hill, in Chili Gulch. Ownership: Neilsen Sand and Gravel Co., c/o C. W. Neilsen, San Andreas; Ray Neilsen, managers

This property was first worked sometime prior to 1900, when a large pit on the southeast side of Chili Gulch was excavated by hydraulicking, and a great mass of tailings was deposited in the gulch. The Neilsens reworked these tailings for a few months in early 1936 with a small 37-bucket dredge. About 60,000 yards of gravel were washed in a 4- by 12-foot trommel and sluices, and yielded between 10 and 15 cents per yard. Since 1939, the Neilsens have produced sand and gravel almost continuously from this property (see section on Sand and Gravel). A sluice with a quicksilver trap was installed in the aggregate plant late in 1957 to recover fine gold.

The gravel consists of virgin gravel of the Chili Gulch channel and hydraulic tailings piles on the east slope of Chili Gulch, and Recent stream gravel in the bottom of the gulch. The tailings, which were accumulated from the Tertiary stream channel, have a notably higher vein-quartz content than the Recent gravels.

The old hydraulic pit east of the aggregate plant was being prepared late in 1957 for mining the following Spring. Use of this virgin gravel was expected to result in an increased yield of gold.

Rough Diamond mine

Location: Portions of SE1/4 sec. 24, T. 5 N., R. 11 E., M.D.M., about 2 miles south of Mokelumne Hill. It is south of and adjacent to the Green Mountain drift mine. Ownership: Raymond and Marv J. Garamendi, Mokelumne Hill, California.

Prior to 1877 this property was worked as a hydraulic mine, but cemented gravel prevented recovery of much of the gold. A Mr. Garland purchased the ground in 1877, drove a 450-foot drift, and erected a five-stamp mill that handled 90 tons of gravel per day. The following year it was bought by Mr. Jillson, and during the next 4 years three small adits were driven, and the main adit was extended to 800 feet, from which points drifts were run 175 feet north and 350 feet south. Gold in the gravel is reported to have averaged $2.00 to $2.50 per carload in 1883, No further activity is recorded until 1934, when the property yielded a small amount of gold. It was inactive again until mid-1942, when exploration for quartz crystals began. Between September 1942 and November 1943, about 6600 pounds of crystal were recovered. Production of modest amounts of quartz crystal continued through 1944. During the latter part of that year, both the Rough Diamond and Calaveras Crystal properties were operated by the Quartz Crystal Products Company. No crystal production has been reported since 1944.

Statements regarding the earliest quartz-crystal production from the Rough Diamond mine (Julihn, and Horton, 1938, p. 71) are not specific, and merely imply production prior to 1938. The name of the mine, which probably refers to quartz crystals found in it, was entered in the patent records in the late 1800s.

This property is on the Tunnel Ridge channel. Its geology is similar to that of the Green Mountain mine and is described under that heading. Reserves of quartz crystal in the Rough Diamond were estimated in 1944 to be 12,600 pounds in 9000 cubic yards of gravel. About 5 percent of this quartz would be of piezoelectric grade (Durrell, 1944, p. 432-433). A later report (Wright, 1950, p. 107) describes this deposit as "nearly exhausted". Development of the Rough Diamond includes, besides the drifts of the 1870s and 1880s, about 1000 feet of drifts trending east and northeast, and a total of 43 feet of raises, driven in 1942-43.

Explosives were used in opening the Rough Diamond Mine, but this practice was discontinued when it was found that crystals were being fractured, and pneumatic tools were used instead. As the result of pilot-plant tests conducted on the property from June to November 1944, a washing plant was erected on the Calaveras crystal (Green Mountain) property. This plant had a rated capacity of about 200 tons of gravel per hour, and consisted of a 35-foot trommel and sluices to recover gold; crystal would be recovered on a conveyor belt and in the pit. It was proposed to reach the crystal-bearing gravel with an open cut of 50-feet bottom width, using a bulldozer and power shovel.

What Cheer mine

Location: NW1/4 sec. 24, T. 5 N., R. 11 E., M.D.M., just north of the Buffalo drift mine, 2 miles southwest of Mokelumne Hill. Ownership: Theresa Murphy and Marv Garamendi, Mokelumne Hill, California.

This property was active prior to 1871, although nothing has been recorded regarding the nature of the work. An inclined adit and several drifts and crosscuts were driven during 1871-72, and a bedrock adit was driven in 1885. In 1887, a five-stamp mill was erected. Shortly afterward, the mine was closed; it remained idle, except for some activity in 1899-1902 and development in 1935-37.

The Deep Blue Lead trends due north of the What Cheer property. Gravel in the channel consists of 20 feet of slightly cemented well-rounded quartz pebbles and boulders on a slate bedrock, capped by 150 feet of volcanic ash; sand, clay, and pyrrhotite arc common. Gold is fine-grained and approximately 900 fine. In 1937 gravel yielded $2 per ton in gold.

The latest development of this property was by a 285-foot vertical shift situated 1300 feet south of an old 600-foot west-trending inclined adit. From the bottom of the vertical shaft, an 1100-foot timbered haulage drift was run north on the east rim of the channel; gravel was mined by inclined crosscuts to the trough of the channel. Between 40 and 50 acres of productive ground reportedly are still virgin (Averill, 1946, p. 254).

Gravel was treated in 1937 by a 14-foot scrubber-trommel, 50 feet of sluices, and a rake classifier. The capacity of this plant was 100 tons per day. [5]

Mountain Ranch

Location. This is a small district in central Calaveras County in the vicinity of the town of Mountain Ranch. It includes the Cave City area.

Geology. The district is largely underlain by graphite schist, quartzite, limestone, and thin bodies of talcose schist. Patches of Tertiary auriferous gravels overlie the bedrock. A few narrow gold-quartz veins, in schist, contain small but sometimes rich ore shoots. The ore contains free gold and abundant sulfides, especially galena.

Mines. Placer: Cotton Flat, Foley, Hidden Cave [Secs. 10 & 11, T. 4 N., R. 13 E.], Humboldt, Mountain Ranch, Rose Hill [Sec. 8, T. 4 N., R. 13 E.].

Murphys

Location. The Murphys district is in south-central Calaveras County about seven miles northeast of Angels Camp. It extends west to include the Esmeralda and Fricot Ranch areas.

History. The streams were first mined during the gold rush. Murphys was established in 1848 or 1849 and named for John M. Murphy, a member of Captain Weber's Stockton Mining Company. Lode mining probably began shortly afterward and continued almost steadily until around World War I. Some of the mines have been prospected in recent years. The old town of Murphys, one of the best-preserved mining towns in the Sierra Nevada, is a popular tourist attraction.

Geology. The eastern part of the district is underlain by graphite schist, slate, and a large limestone lens of the Calaveras Formation (Carboniferous to Permian). The western part is underlain by slate, schist, green schist and numerous lenses of talcose rock derived from serpentine. A number of patches of auriferous Tertiary gravels overlie the bedrock.

Ore Deposits: Quartz, occurring in a great many west-trending veins—mostly in schist and slate—is glassy and white, rose, or occasionally black. Rosecolored quartz is characteristic of this district. The ore bodies usually are small and shallow, but often they are rich. The ore contains free gold and often abundant sulfides. Although there are no large mines, the district was quite productive because of the large number of veins.

Mines. Basco, Beatrice, Bence $200,000, Bonchard, Buckeye, Buckhorn, Crown Point, Cowbell, Dora Cons., Dragone, Economic, Esmeralda $300,000, Eureka, Fairplay, Falcon, Fricot Group, Great Divide, Gumboot, Hidden Treasure, K and J, Last Chance, Malteson, Manhattan, Miralda, Oro y Plata, Piety Hill, Rocky Bar, Total Wreck.

Railroad Flat

Location. This district is in central Calaveras County in the vicinity of the town of Railroad Flat, seven miles south of West Point and 13 miles cast of Mokelumne Hill. The district was named by an early day placer miner who had laid a few hundred feet of wooden railroad track on his claim.

Geology. The area is underlain chiefly by graphitic schist, slate, quartzite, and metamorphosed chert. A number of narrow north-striking quartz veins contain free gold and often abundant sulfides, especially galena. The ore shoots are small and usually do not extend to depths of greater than 200 feet, but they often are rich.

Several deposits of quartz-rich gravels have been mined by drifting or hydraulicking. These gravels were deposited by the south-trending Tertiary Fort Mountain channel. In places the gravels are overlain by rhyolite.

Mines. A.V.G., Bald Eagle, Banner Blue (placer), Clary, Fine Gold $200,000, Jeff Davis,

Kaiser Wilhelm, Lampson (placer), Mohawk, Old Gray, Petticoat, Poe, Prussian Hill, Sanderson $100,000+, Summit, Swiss.

Rich Gulch

Location. Rich Gulch is in north-central Calaveras County adjacent to the Mokelumne Hill district on the east and west of the West Point and Railroad Flat districts. It includes the Jesus Maria area.

Geology and Ore Deposits. The area is underlain by graphite schist, quartzite, slate and numerous limestone lenses of the Calaveras Formation (Carboniferous to Permian) that have been intruded by several small granodiorite stocks.. The quartz veins usually are narrow and contain varying amounts of gold and sulfides. Much high-grade ore was recovered from shallow workings during the early days, but at depth the ore usually contains only a few dollars per ton in gold.

Mines. Blue Jay, Ilex, Quartz Glen $300,000+, Rindge No. 1, No. 2, No. 3, Salvador.

San Andreas

Location. This district is in west-central Calaveras County. It consists of that part of the Mother Lode gold belt that extends from the North Branch-Cottage Springs area southeast through San Andreas and Kentucky House to the vicinity of Fourth Crossing. It is both a lode- and placer-mining district.

History. The streams were mined during the early part of the gold rush. San Andreas Gulch was first settled by Mexicans in 1848 or 1849. The Tertiary channel deposits were mined by hydraulicking and drifting from the 1850s through the 1880s, and some drift mining has continued until the present time. Lode mining also began in the 1850s. Numerous lode mines were active from the 1870s until about 1900 and again during the 1930s. The Union gold mine was prospected for uranium in the middle 1950s.

Geology. The area is underlain by northwest trending belts of greenstone, amphibolite schist, slate, and lenticular serpentine bodies. At Kentucky House, a large body of dolomitic limestone contains lenses of high-calcium limestone mined by the Calaveras Cement Company. To the east of San Andreas is a granodiorite stock. Overlying portions of the district are numerous deposits of auriferous Tertiary channel gravels.

Ore Deposits. The lode deposits consist of northwest-trending quartz veins with stringers that contain finely divided free gold and pyrite. The veins occur principally in amphibolite and greenstone. Most of the milling ore averages 1/3 ounce of gold per ton or less, but several of the ore shoots have been mined to depths of 700 feet. Considerable high-grade ore was taken from shallow workings during the early days. Small amounts of tellurides and uranium minerals have been found in this district.

Mines. Placer: Benson, Central Hill, Lloyd, North Branch, Rising Star, Wheats.

• Lloyd (Watson Bros.) mine

Location: SE1/4 sec. 2 and NE1/4 sec. 11, T. 4 N., R. 11 E., M.D.M., 3 miles northeast of San Andreas, east of Latimer Gulch. Ownership: Lottie C. Jackson, San Andreas, California.

This drift mine on the Central Hill channel was described in 1873 as "two-fifths drifted out" through a 44-foot shaft. An 800-foot adit was used to drain the mine. An arrastre treated 12 tons of gravel per day. A 5-stamp mill on the property, which formerly crushed gravel, was used for custom mining. The Lloyd mine was again active in the early 1900s and during 1932-39.

The channel contains 8 feet of pay gravel consisting of schist, slate, and quartz. It rests on blue slate and is capped successively by 2 feet of volcanic mud, 20 feet of ash, a sequence of mixed ash and gravel, and 50 feet of sand and gravel. A younger cross-channel north of the inclined shaft trends N. 20° W. It contains diorite and quartz cobbles and has cut down below the Central Hill channel. The bedrock of the Central Hill channel pitches gently south, but water flowed north during the Tertiary period. In 1934 some 12,000 tons of gravel yielded 802 ounces of gold and 124 ounces of silver. Gold is 875 fine; its size ranges from "pin-head" to "wheat-grain". In the 1930s 1-1/2 feet of bedrock and 5 feet of gravel were mined.

Workings consist of a 300-foot adit inclined 33° E., a 750-foot drift north from the shaft station, a 200-foot drift northwest on the younger channel, drifts to the east from the shaft station and north workings, and breasting west of the main (north) drift. In 1937 a vertical shaft 100 feet north of the incline was sunk to a depth of about 85 feet, and gravel was drifted to the east and south. During the 1930s gravel was treated in a 4- by 14-foot scrubber-trommel, sluices, a rake classifier, and a cone classifier.

Rising Star (Cassinelli) mine

Location: Part of NE1/4 SE1/4, and part of W1/2 SE1/4, sec. 2, T. 4 N., R. 11 E., M.D.M., about 3 miles northwest of San Andreas, just east of Latimer Gulch. Ownership: Roy and Edna Byrns, 73 Los Banos Avenue, Daly City, California.

Gravel of the Deep Blue Lead was first encountered on this property in 1871 after a Mr. Mullen had driven an 1100-foot adit to a point 30 feet above bedrock. Subsequent drifting outlined the lateral extent of the pay gravel. It was reported in 1875 that a month's work yielded 160 ounces of gold. In 1892 the property was being prospected by a 1200-foot tunnel. The following year workings consisted of two shafts and a 1500-foot adit, which by 1900 had been extended an additional 300 feet. The gravel was estimated to contain $4 per ton in gold in 1896. The Rising Star mine was intermittently active from 1900 through 1916. Work during this period consisted of breasting the gravel, which was then treated by slacking and washing in a trommel and sluices. During 1920-25 and 1932-39 the mine was again prospected and developed. Work during the 1930s was south of the main adit on a channel known as the Blue Wash. A drift was run southward for 500 feet from the old Gleason workings in order to explore the cast rim of this channel.

The Rising Star property lies across the Deep Blue Lead, which here consists of four separate channels that flowed slightly east of south over a relatively flat bedrock basin. From east to west these channels are the Lloyd, Mullen, Gleason, and Blue Wash. The basin occupied by these channels is 1200 feet wide. The gravel in all four channels consists of a heavy wash containing blue clay and many large boulders and cobbles. It is compact, sometimes cemented, and generally about 30 feet deep and 100 feet wide. It rests on a bedrock of slate, schist, and gabbro, and is capped by a breccia of andesite and rhyolite pebbles overlain by surface gravel. Gold ranges in size from medium to coarse and is found with considerable amounts of black sand and pyrite. It is chiefly in the upper 2 feet of bedrock and the lower 3 feet of gravel.

Workings in the Rising Star mine include two main ventilation shafts, the Mullen and Gleason, and more than 3000 feet of drifts and crosscuts. The entry adit of the main tunnel is situated 200 feet cast of the east rim. The tunnel trends southwest for 800 feet, crossing the Lloyd channel and penetrating the Mullen channel. It is connected with the Mullen shaft, beyond which it trends west for 700 feet, cutting the Gleason channel. Near the center of the Blue Wash the main tunnel swings southward and continues down-channel for about 650 feet. A series of crosscuts run to the west during the 1930s in this portion of the tunnel were said to have disclosed 100 feet of pay gravel. Blue Wash gravel was mined in the old days by breasting for 300 feet north of the main runnel between the Foley and Gleason shafts. Another wide block of breasted gravel is in the Gleason channel and is the means by which the Gleason shaft was connected with the main tunnel 200 feet to the south. The Mullen channel was breasted south of the tunnel during the early life of the property. The Lloyd channel has not been worked from the main tunnel.

During the 1930s, gravel was treated in a 3- by 12-foot trommel and about 70 feet of sluices. [5]

Sheep Ranch

Location. The Sheep Ranch district is in south-central Calaveras County about 16 miles east of San Andreas and eight miles north of Murphys. It includes the old Washington district just to the south and the El Dorado area to the west.

History. The streams in the area were first mined during the gold rush. Drift mining of the Tertiary channel gravels began in the late 1850s and continued intermittently through the early 1900s. The Sheep Ranch mine, the largest source of gold in the district and the largest mine of the Sierra Nevada east gold belt, was first worked in 1868. Senator George Hearst had an interest in this mine from 1877 to 1895. It was operated on a major scale until around 1907 and again from 1936 to 1942. The Right Bower mine has been intermittently worked since 1946.

Geology. The principal rocks underlying the area are slate, impure quartzite and graphite schist of the Calaveras Formation (Carboniferous to Permian), which have been intruded by gibbroic stocks. The gravel deposits are quartz-rich and occur as patches. They are part of the Tertiary Fort Mountain channel, which extends south from the Railroad Flat district. In places the gravels are overlain by rhyolite.

Ore Deposits. The quartz veins usually are a few feet thick and range from white to dark smoky gray in color. Smoky vein quartz is characteristic of some of the mines in this district. The ore contains free gold, which often is coarse, and varying amounts of sulfides. Appreciable quantities of high-grade ore have been recovered. The Sheep Ranch vein has been mined to a depth of about 3000 feet. The gravels usually are fairly well-cemented and were rich in places.

Vallecito

Location. Vallecito is in south-central Calaveras County about five miles east of Angels Camp and five miles south of Murphys. It is a placer-mining district and includes the Douglas Flat and Dead Horse Flat areas.

History. Daniel and John Murphy found gold here in 1848, and the district was first known as Murphys' Old Diggins. The name was changed to Vallecito in 1854. The hydraulic mines were active from the late 1850s through the 1880s. Drift mining was done at this time and again during the 1930s, when the Vallecito Western drift mine was operated on a fairly large scale. Dragline dredging was active in the district during the 1930s.

Geology. The Tertiary Central Hill channel enters the district from the north from Murphys and then extends west toward Altaville and Angels' Camp. Here the channel is joined by two small tributaries from the east and south, the one from the east known as the Murphy's Gulch channel. Farther east is the south trending Cataract channel. The gravels, consisting of granitic material and quartz, were richest near bedrock. The gold was fairly coarse. The gravels are overlain in places by rhyolite and andesite and also by some Pleistocene gravels. Bedrock is limestone, schist, slate, and amphibolite, and to the west there is granodiorite.

• Golden River mine

Location: SW1/4 sec. 19, T. 3 N., R. 14 E., M.D.M., three-fourths of a mile northwest of Vallecito. Ownership: T. F. and F. K. Stearman, Vallecito, California.

The Golden River Mining Company, organized by Roland R. Woolley in 1933, acquired several leases on property in the Vallecito area from the Thomas Bishop Company that year and surrendered them in 1950. This property was subsequently sold in parcels to various new owners. The Golden River drift mine was active from 1933 to 1938, during which time more than 5700 ounces of gold were recovered (Roland Woolley; personal communication, 1957). The character of the gravel is the same although the gold is much coarser than in the Vallecito-Western mine, which is adjacent downstream. Nuggets as much as 2 inches in greatest dimension were common. Gold is in crevices in the bedrock and in partly cemented pay gravel. The pay gravel, which is from 40 to 150 feet wide and from 4 to 10 feet thick, contains many boulders. It rests on a dominantly granitic bedrock and is disrupted by extensive faulting that has caused vertical offsets of bedrock as much as 16 feet in height. To avoid these offsets, an 1800-foot main haulageway was driven in bedrock beneath the channel, and raises were put up to intersect the gravel at various points. The gravel is capped by rhyolite ruff.

Development includes a 310-foot vertical shaft, more than 2000 feet of drifts upstream (east), and 1000 feet of drifts downstream. In 1937, a second shaft, 232 feet deep, was sunk 1 mile east of the first shaft. Gravel was washed in a 40-foot scrubber-trommel and 300 feet of sluices. About 90 percent of the gold was recovered in the first 10 feet.

From 1935 to 1938, subleases were let on portions of the surface near Vallecito, and dredging operations there recovered an estimated $200,000 above the amount credited to the mine (Roland Woolley, personal communication, 1957). In 1958, several shallow drill holes sunk by Joseph Mancuso Jr., a lessee from Mr. Stearman, revealed a modest amount of medium-coarse to coarse gold in surface gravel (Ted Fisher, personal communication, 1958).

Vallecito Western mine

Location: SW1/4 sec. 24, T. 3 N., R. 13 E., M.D.M., near Six Mile Creek, 2 miles west-northwest of Vallecito. Ownership: Thomas B. Bishop Co., 166 Geary Street, San Francisco, California.

In 1923 Donald Steffa and three other men ceased their activities on the Vallecito Consolidated property at the north end of the town of Vallecito and began prospecting the Central Hill channel in the vicinity of Six Mile Creek. They organized the Vallecito Mining Company, obtained a lease on 285 acres of property from the Thomas Bishop Company, and began drilling in May 1923. By May 1924 seven holes totaling about 2000 feet had been drilled. From the information thus obtained, a cross-section of the channel was made, and a site selected on which a 4- by 71/2-foot shaft was sunk to a depth of 167 feet. The property was developed by 6300 feet of workings by the end of July 1931.

The Vallecito Western yielded 4,206 ounces of gold by 1933. From December 1932 to October 1936, the mine was operated under a sublease by the Tonopah-Belmont Development Company. From January 1933 to September 1936, 40,967 tons of gravel yielded

7,222 ounces of gold and 815 ounces of silver. In March 1935, a major fault was encountered in the main gangway about 3000 feet northeast of the shaft. The channel was cut off completely, and was not found again until the following December. The Vallecito Mining Company operated the mine in 1937 and recovered 358 ounces of gold from 1,213 tons of gravel. In 1938, 917 tons of gravel were mined, and 27,223 yards of old tailings were washed. A dragline dredge was operated on the property during 1939, and a sub-lessee recovered 274 ounces of gold and 31 ounces of silver from 687 yards of gravel obtained by drift mining. Vallecito Mining Company worked the mine again throughout 1940 and recovered 221 ounces of gold and 25 ounces of silver from 685 yards of gravel. Activity ceased on the property in 1941, and has not resumed.

The Vallecito Western mine is situated on a 275-foot-wide channel in a Tertiary valley that is approximately 1400 feet wide. A major fault completely offsets the channel about 3000 feet northeast of the shaft. Bedrock slate, amphibolite, and granodiorite are overlain by 15 to 70 feet of compact, uncemented gravel. The gravel is well-rounded and contains many large boulders in a matrix of cobbles and fine sand. Cobbles and boulders are composed chiefly of granitic rock; small proportions of slate, quartzite, jasper, and quartz also are present. Overburden consisting of sand, gravel, and rhyolitic ash ranges from 100 to 250 feet in thickness.

Pay streaks are concentrated in the main trough of the channel and on its south rim. Breasting was carried across the full width of the pay streak, which ranges from 40 to 150 feet. Pay streaks are uniform over long distances and are accompanied by only small amounts of black sand. About 75 percent of the gold is on, or within, 1 foot of bedrock, although pay streaks have been found as high as 23 feet above bedrock. The average breast height was 6 feet. Where there is gold well above the bedrock, there is usually very little near the base of the gravel.

The gold is in coarse, flat flakes up to 5 ounces in weight; about 90 percent of it is coarser than 20 mesh. The average size of the gold increased as workings were advanced upstream. Recovery of small amounts of crystalline and wire gold was reported in 1929. The fineness of the gold ranged from 882 to 900, and averaged 895.

The mine was developed by a 167-foot vertical shaft that includes a 14-foot sump below the shaft station, and more than 6300 feet of drifts and crosscuts. From the shaft, a diagonal crosscut was run southeast to a point near the north rim of the channel, and a drift was then run east parallel to the rim. Another diagonal crosscut was run southeast, beginning 250 feet from the shaft and extending to a point near the south rim of the channel, from which another drift was advanced to the east. Crosscuts between the parallel drifts were run at intervals of 100 to 150 feet, and other crosscuts were projected away from the drifts onto benches and up the rims.

Gravel was washed in a 3- by 18-foot scrubber-trommel, after passing through an 101-foot sluice where large cobbles were removed by hand. The trommel portion consisted of two concentric tubes of plate punched with 1-1/2-inch holes on the inner tube, and 3/8-inch holes on the outer tube. Oversize gravel was passed through a sluice and over riffles. Undersize (minus 3/8-inch) was passed over riffles, onto a table, and over more riffles. Steffa reported in 1932 that 200 tons of re-washed tailings yielded only 17-1/2 cents per ton; the gravel had originally contained $8.00 per ton. [5]

West Point

Location. This extensive Sierra Nevada east gold belt district is in eastern Amador and Calaveras Counties in the general area of the town of West Point. It includes the Skull Flat, Glencoe, Bummerville, pioneer Station, and Buckhorn areas.

History. The town was first known as Indian Gulch but was renamed West Point after a geographic feature discovered by Kit Carson while he was enroute to Sutter's Fort in 1844. The streams and surface ores were mined extensively during the 1850s, when large amounts of gold were recovered. During the 1860s and 1870s many lode mines and 10 or more custom mills were active, but there was much difficulty with sulfides. Some activity was noted from the 1880s until 1914 and again during the 1920s and 1930s. Several mines have been intermittently worked since World War II, the chief operations having been at the Belden, Blackstone, and Centennial mines. This is one of the more productive districts of the east gold belt, and an extremely large number of mines exist.

Geology. The gold deposits are associated with a west-elongated body of granodiorite five miles wide and 15 miles long that has intruded graphitic slates, quartzites, and schists of the Calaveras Formation (Carboniferous to Permian).

Ore Deposits. Numerous north-trending and west dipping (a few dip east) quartz veins are found in the granodiorite or in the adjacent metamorphic rocks. The veins usually are one to five feet thick, have persistent strikes, and belong to one of three main vein systems that have not been mapped. Narrow diorite, quartz-diorite, and aphlite dikes commonly are associated with the veins. The ore bodies contain free gold and abundant sulfides, especially galena, which is nearly always associated with high-grade ore. The ore shoots usually have horizontal stoping lengths of 150 feet or less, but several were 300 to 400 feet long. Milling-grade ore commonly averages one ounce or more in gold per ton, and much high-grade ore has been recovered. Few of the mines have been developed to depths of more than a few hundred feet. It has been estimated that there are more than 500 mine shafts in the district.

Mines. Amador County: Amador-Columbus, Belden $400,000+, Black Prince $100,000+, Defender $100,000+, Elkhorn, Hageman, Jumbo, Lone Willow $100,000+, Newman $160,000+, Pine Grove, Pioneer Lucky Strike $300,000+, T.N.T. Calaveras County: Austrian, Billy Williams, Backstone $200,000, Blazing Star, Buena Vista, Carlton, Centennial, Champion $500,000, Chino, Continental $100,000+, Corn Meal, Cross, Etna, Ever Ready, Fidelity, Garibaldi, Gilded Age, Glencoe, Golden Rule, Gold Star, Good Hope, Keltz $300,000+, Lockwood $400,000+, Lone Star, Marquis, Mina Rica, Monte Cristo, North Star, Old Henry, Rindge No. 1, 2, and 3, Riverside, San Bruno, San Pedro, Scorpion, Soap Root, Star of the West, Swallow, Water Lily, Wide West, Woodhouse $100,000+, Yellow Aster $100,000+.

Gold Dredging in Calaveras County

• Calaveras County has been famous for its surface placers, especially during the days of the "long-tom" and "rocker." The gold now produced comes mostly from deep quartz and gravel mines. Gold dredging has only been carried on since 1904, and so far only in two localities. At the present time, 1910, there are three dredging companies in active operation—The Mokelumne Mining Company, Calaveras Gold Dredging Company, and Isabel Dredging Company. The Butte Dredging Company of Oroville is this year removing the Butte dredge to property adjoining that of the Isabel Dredging Company on the south side of the Calaveras Rivcr.

The Mokelumne Mining Company was incorporated under the laws of Delaware in 1901. The president and general manager of the company is David Pepper, Jr., Philadelphia, Pa., and the resident manager is William C. Calley, Calaveras County, Cal.

The holdings of the company comprise an area of about 250 acres of dredging land located along the Mokelumne River in Calaveras, San Joaquin, and Amador counties at the junction of those three counties, about three miles from the town of Wallace.

The Calaveras Gold Dredging Company began operations in 1904. The president and manager of this company is S. A. Moss, 807 Alaska Commercial Building, San Francisco, Cal. The holdings of this company comprise an area of about 350 acres located in township 3 north, range 10 east, along the Calaveras River near Jenny Lind. The ground averages in depth about 33 feet. The gravel is coarse and in most places overlain by hydraulic tailing. [2]

EL DORADO COUNTY

Coloma

Location. This district is in the vicinity of the old mining town of Coloma in western El Dorado County. It is on the South Fork of the American River about eight miles northwest of Placerville.

History. Although this is a relatively small district, it is significant, for it was here that James W. Marshall made his historic gold discovery. In August 1847, Captain John Sutter, grantee of a large Mexican land grant in the vicinity of present-day Sacramento, signed a contract with Marshall to erect a sawmill on the American River. Work commenced in September 1847. The mill was almost finished on January 24, 1848, when Marshall, inspecting the mill tailrace, noted several small flakes of what appeared to be gold. Work on the mill stopped, and more flakes were recovered. These were taken to Sutter at Sacramento for more tests, which proved beyond a doubt that it was gold. Attempts were made to keep the discovery a secret, but the news quickly leaked out.

Soon Sonorans from the Los Angeles placer-mining districts arrived, the vanguard of the thousands of gold seekers who came from all directions to Coloma. The surface placers here were soon exhausted, and the miners went elsewhere. Coloma was named for a nearby Southern Maidu Indian Village. Early spellings were "Colluma" and "Culoma". Some gold dredging was done on the American River here during the 1930s and 1940s.

Marshall never was associated with a really successful mining venture and died in 1885 in the nearby town of Kelsey, a poor man. The Marshall Monument, where he is buried, was dedicated in 1890. Part of the old town, the mill site and the monument joined the California state park system in 1927. A replica of Sutter's mill was recently constructed at the park. Also at the park is a museum containing many items of early-day mining equipment.

Geology and ore deposits. The central portion of the district is underlain by a granodiorite intrusion. It is surrounded by slate, mica schist, amphibolite, and several north-trending lenticular bodies of serpentine.

Most of the gold values were obtained from gravels in the American River or from terrace gravels along the bank. A few narrow gold-quartz veins crop out and several contact-metamorphic copper-gold deposits are found along the margin of the granodiorite.

Fairplay

Location and History. This district is in south-central El Dorado County about 20 miles southeast of Placerville. It includes the Slug Gulch and Cedarville areas. It is primarily a placer-gold district, but some copper has been mined here. The area was first settled in 1853 by N. Sisson and Charles Staples. The name, according to tradition, arose from an incident in which an appeal for fair play forestalled a fight between two miners.

Geology. The deposits here are part of an isolated Tertiary gravel channel that extends southwest from Slug Gulch toward Fairplay. The gravels are of various ages. Older benches of quartz gravel are rich, younger intervolcanic gravels are leaner. The gold is extremely coarse. At Slug Gulch the bedrock is limestone; elsewhere it is slate and schist; to the west is granodiorite. A few narrow gold-quartz veins are found in the district.

Georgetown

Location. The Georgetown district is in northwestern El Dorado County at the north end of the northeast segment of the Mother Lode belt. It extends from just north of Garden Valley north through Georgetown and the Georgia Slide area to the Middle Fork of the American River. It is both a lode- and placer-mining district.

History. Mining began here in 1849 by a party of placer miners from Oregon. The site was first known as Growlersburg, but was soon changed to Georgetown. It is reported to have been named for either George Ehrenhaft, who laid out the town, or George Phipps, a sailor-prospector. The placers were highly productive during the 1850s. The Seam deposits at Georgia Slide were mined on a large scale by hydraulicking from 1853 to about 1895. There was some activity during the early 1900s, and in the 1930s the Beebe and Alpine mines were worked on a fairly large scale. There has been minor prospecting and skin diving in the district since.

Geology. This district is in the northern end of the Mother Lode gold belt. There is a two-mile wide north- and northeast-trending belt of Mariposa slate (Upper Jurassic) in the central portion of the district, with greenstone and green schist to the west and mica schist, slate, quartzite, amphibolite, and serpentine to the east. In places, especially at Georgia Slide, the bedrock is deeply weathered. Several patches of Tertiary gravel overlain by andesite are exposed on some hills in the northern part.

Ore Deposits. Ore deposits consist of thick zones of mineralized schist and slate that contain numerous quartz veins and veinlets. Where deeply weathered the gold became concentrated and such deposits were worked by placer-mining methods. These are known as "seam" deposits. Below the weathered zone they were mined as lode deposits. The seam deposit at Georgia Slide was 1000 feet long and 500 feet wide. Usually the milling ore yielded from 1/7 to 1/5 ounce to the ton, but many high-grade pockets were encountered. In addition, there are several wide quartz veins containing finely disseminated free gold and pyrite. These veins contained ore shoots with stoping lengths of up to 500 feet. The Tertiary gravel patches have yielded gold.

Mines. Placer: Anderson, Bottle Hill, Cary, Cement Hill, Holmes, Jones Hill, Little Chief, Mulvey Point, Patterson, Rowe, Shoemaker, Tanksley, Trimble.

Greenwood

Location. This district is in northwestern El Dorado County. It consists of the northwest segment of the Mother Lode gold belt, which splits at Garden Valley (the east segment continues through Georgetown). This segment of the belt is several miles wide and extends from the point of the split northwest through Greenwood and Spanish Dry Diggings to the Middle Fork of the American River, a distance of about eight miles.

History. Placer mining began in this area shortly after the beginning of the gold rush. The town was named for Caleb and John Greenwood, who established a trading post here in 1850. The district flourished during the 1850s, when the American River was mined and the seam deposits hydraulicked on a large scale. The river was mined by diverting the main stream with a series of flumes, tunnels, and wingdams. The gold-bearing gravels were removed from the bedrock and sent through sluices or long toms. Major mining activity continued through the early 1900s, much of the later placer mining done by Chinese. This district was quite productive again during the 1930s, when the Sliger, Taylor, and Grit lode mines were active. Since about 1955, numerous skin divers have been mining the Middle Fork of the American River by small-scale methods.

Geology. There are two northwest-trending belts of slate of the Mariposa Formation (Upper

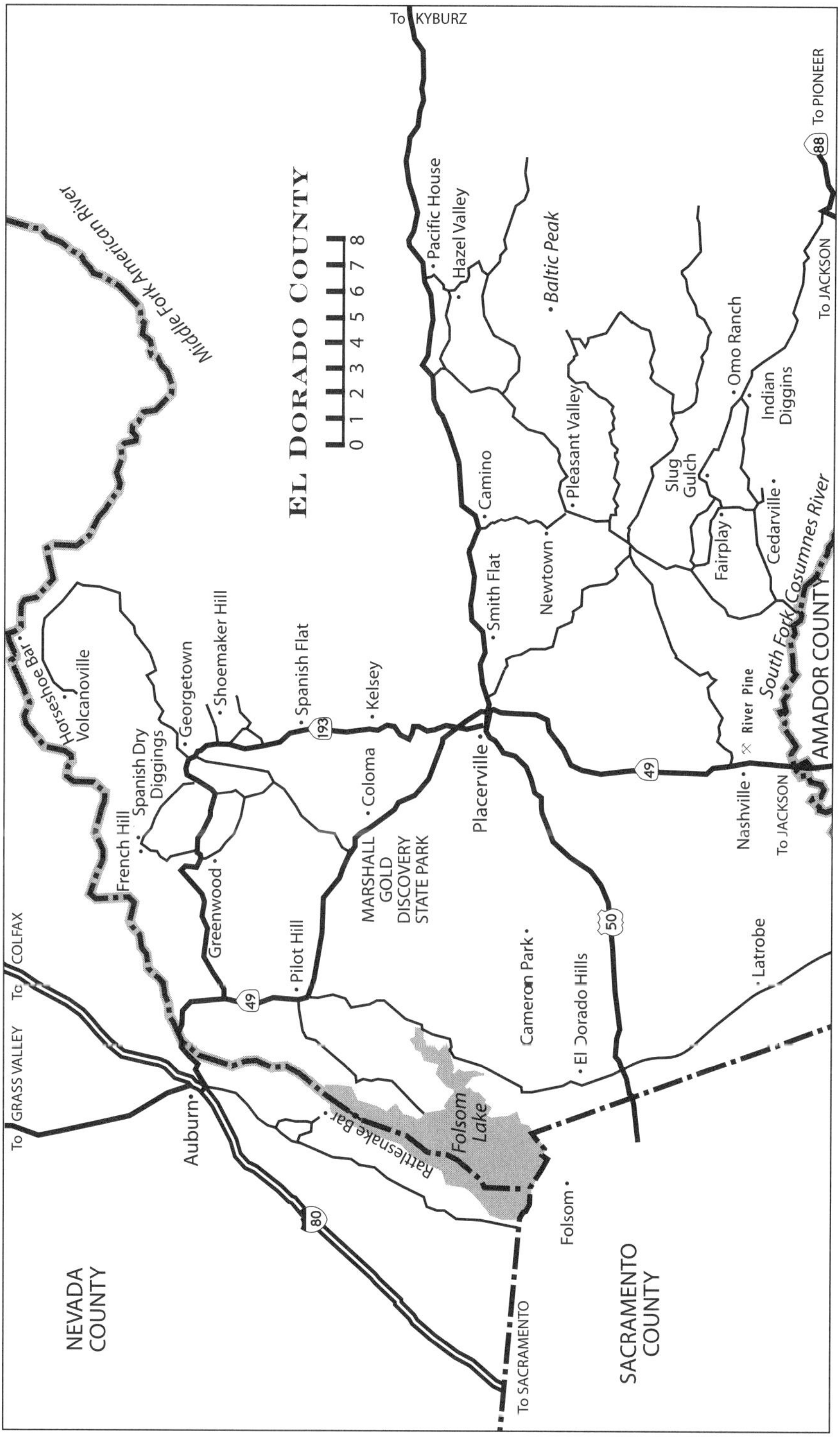
EL DORADO COUNTY
0 1 2 3 4 5 6 7 8
To KYBURZ
To PIONEER
88
To JACKSON
Pacific House
Hazel Valley
Baltic Peak
Middle Fork American River
Omo Ranch
Indian Diggins
Pleasant Valley
Camino
Slug Gulch
Cedarville
Fairplay
Newtown
Smith Flat
South Fork Cosumnes River
AMADOR COUNTY
River Pine
Horseshoe Bar
Volcanoville
Shoemaker Hill
Georgetown
Spanish Flat
Kelsey
193
Spanish Dry Diggings
Coloma
Placerville
49
Nashville
To JACKSON
French Hill
Greenwood
MARSHALL GOLD DISCOVERY STATE PARK
50
To COLFAX
Pilot Hill
Cameron Park
El Dorado Hills
Latrobe
To GRASS VALLEY
Auburn
Rattlesnake Bar
Folsom Lake
80
Folsom
NEVADA COUNTY
SACRAMENTO COUNTY
To SACRAMENTO

Jurassic) 1/2 to one mile apart. Chert, impure quartzite, and slate lie to the west, and greenstone and amphibolite schist lie in the center and to the cast. A number of small lenticular bodies of serpentine and talcose schist are enclosed in the bedrock, which is deeply weathered in places.

Ore Deposits. There are several wide and sometimes extensive zones of quartz veins and veinlets and mineralized schist containing free gold and auriferous pyrite. Where deeply weathered, the bedrock was eroded, and the gold in the seams and veinlets remained and became concentrated. Such deposits are known as seam deposits or "seam diggings". The upper portions were mined by hydraulicking, and later the unweathered veins at depth were mined by conventional underground methods. Considerable specimen material has been recovered from this district, including crystallized gold. The famous Fricot nugget of crystallized gold (201 ounces), which was taken from the Grit mine in 1865, is on display in the Division of Mines and Geology exhibit in the Ferry Building, San Francisco. Milling-grade ore bodies commonly averaged 1/5 to more than 1/2 ounce of gold per ton. Some of the veins were mined to inclined depths of 2000 feet.

Mines. Admiral Schley, Argonaut $100,000+, Bazocoo, Cedarburg, Centennial, Esperanza $100,000+, Eagle, French Hill $100,000+?, Greenwood, Grit, Hines-Gilbert $100,000+, Homestake, Maltby, Nancy Lee, Oakland Cons., Railroad Hill, Red Mount, Revenge, Rosecranz $100,000+, San Martin, Sebastopol, Sliger $2.85 million, Taylor $1 million.

Grizzly Flat

Location. The Grizzly Flat district is in south-central El Dorado County about 25 miles east of Placerville. It is in the Sierra Nevada east gold belt and includes the Hazel Valley and Baltic Peak areas. It is both a lode and placer gold-mining district.

History. The streams were originally mined during the gold rush. The camp was established in 1850 and named for a grizzly bear that surprised a group of miners during an evening meal. The Mt. Pleasant mine was discovered in 1851. There was much activity in both the lode and placer mines from the 1870s through the early 1900s. There was some mining activity again during the 1930s. The Hazel Creek mine was discovered in 1948 and was worked on a fair-sized scale until 1958.

Geology. The southern part of the district is underlain by granodiorite, which extends west from the mass of the Sierra Nevada batholith. To the north is slate, phyllite, and graphitic and mica schist of the Calaveras Formation (Carboniferous to Permian), which in places contains small tactite bodies. There are several patches of Tertiary gravels overlain by andesite that are part of the south-extending Tertiary channel of the Mokelumne River.

Ore Deposits. A number of north-trending quartz veins and stringers are found in both the granodiorite and metamorphic rocks. The ore contains free gold and abundant sulfides, especially galena. The ore averages 1/2 to more than one ounce of gold per ton, but few ore shoots were more than 100 feet long or attained much depth. The Tertiary channel gravels are thin and the channels narrow, but in some places they were extremely rich.

Mines. Placer: Grizzly Flat drift, Payne drift.

Indian Diggings

Location. This district is in south-central El Dorado County about 30 miles southeast of Placerville. It includes the Indian Diggings, Henry Diggings, Omo Ranch and Brownsville areas. Indian Diggings is best known as a placer-mining district, but there are a number of lode deposits. The Fairplay district lies to the west.

Geology. The district is underlain by quartz-mica schist, graphitic slate, green schist, quartzite, and limestone. The central part of the area has been intruded by a round quartz-diorite stock. Portions of the bedrock are overlain by patches of Tertiary auriferous gravel and extensive andesite flows.

Ore Deposits. The channel gravels are part of the Tertiary Mokelumne River, which extends south into this area from the Grizzly Flat district. The channel then extended west and southwest toward Fiddletown in Amador County. Indian Diggings was on a branch of this channel. Large amounts of gold came from these channel deposits, especially those at Indian Diggings, where the bedrock is limestone that contained numerous rich potholes. Mining was done by both hydraulicking and drifting. A number of narrow north-striking quartz veins in quartz diorite contain small but often rich ore shoots. The ore contains free gold and abundant pyrite, galena, and smaller amounts of other sulfides.

Mines. Placer: April Fool, Armstrong, Carrie Hale, Chic, Christion, Deep Channel, Dorsey, Drusy, Hayward, Hidden Treasure, Irish Slide, Last Chance, Little Bill, Lucky Jack, Old Chink, Omo, Patterson, Payne, Peacock, Richmond, Syracuse, Telegraph, Tomcat, Yellow Aster, Yellowjacket.

Kelsey

Location. The Kelsey district is in northwestern El Dorado County. It is that portion of the Mother Lode gold belt that extends from the vicinity of the town of Kelsey northwest to Garden Valley.

History. This area was placer-mined soon after James Marshall's gold discovery in 1848 at Coloma, a few miles to the west. The camp was first settled by and named for Benjamin Kelsey. Marshall spent his last days at Kelsey, and a building on his property once housed a pioneer museum. There was much lode mining from the 1860s through the early 1900s and again in the 1930s. Some intermittent work has been done recently at the Black Oak mine.

Geology and Ore Deposits. A northwest-trending two-mile-wide belt of gray to black slate of the Mariposa Formation (Upper Jurassic) is in the central portion of the district, with greenstone, slate, graphite schist, and quartzite to the west. Amphibolite, slate, and schist lie to the east. Serpentine lenses are also present both to the east and west. The ore deposits occur in quartz veins with numerous stringers. The veins range from one to 10 feet in thickness. Nearly half of the known output of some of the mines has been from small but extremely rich ore shoots. None of the mines has been worked to depths of more than 600 feet.

Mines. Big Four, Big Sandy $100,000+, Black Oak $1.25 million, Gopher Hill, Gray Eagle, Hart, Ida Livingston, Kelsey $100,000+, Lady Emma, St. Clair, Veerkamp, War Eagle, Yuba.

Newtown

Location. Newtown is in central El Dorado County 10 miles east of Placerville. It was originally known as Dogtown. Chiefly a placer-mining district, it includes the Camino and Pleasant Valley areas.

Geology. Numerous deposits of auriferous gravels were deposited in the west-trending Tertiary channel of the South Fork of the American River. They were mined during the early days by hydraulicking and drifting, and again by drifting through the 1930s. The lower gravels are quartz-rich and in places contain coarse gold. They are overlain by rhyolite and andesite. Bedrock is slate, schist, and quartzite with thin bands of serpentine. Granodiorite crops out to the north and south.

Pacific

This is a placer-mining district in east-central El Dorado County, in the vicinity of Pacific House and about 20 miles east of Placerville. Several hydraulic and drift mines here were originally worked in 1850s or 1860s, with some work at the Pacific Channel drift mine in the early 1920s. The gravel deposits are on a Tertiary channel that extended west and southwest toward Placerville. The gravels are capped by andesite and rest on a granite bedrock.

Pilot Hill

Location. This district is in northwestern El Dorado County in the vicinity of the town of Pilot Hill. The area was first mined during the gold rush. It was worked again in the 1910s with draglines and power shovels.

Geology. The principal deposit is a 20- to 30-acre remnant of the main Tertiary channel of the American River that extended northwest from Placerville. The gravels rest on greenstone and green schist. To the south are several bodies of gabbro and diorite. During the 1930s the yield from these gravels was 13 to 60 cents per yard in gold. Also, there are a few narrow gold-quartz veins.

Placerville

Location. Placerville is in west-central El Dorado County. The district includes the lode mines of the Mother Lode belt, which extends north through the district, and the placer deposits here and in the adjacent Smith Flat, Diamond Springs, Texas Hill, Coon Hollow, and White Rock areas.

History. Gold was discovered in the Placerville area in July 1848. The town was first known as Dry Diggings but had the nickname of Old Hangtown; three robbers were hanged here on October 17, 1849. From the middle 1850s through the 1870s, the hydraulic and drift mines in the district were extremely rich. One 20-acre claim at Coon Hollow yielded $5 million, and the Spanish Hill area yielded $6 million. Quartz mining began in 1852 at the Pacific mine, but the chief period of lode mining was from the 1880s until about 1915. There was some mining in the district again in the 1930s, but there has been little activity since. Many of the mines in the district came under the control of the Placerville Gold Mining Company. The value of the total output for the district is unknown, but the placer mines are estimated to have yielded at least $25 million.

Geology. A belt of gray to black slate of the Mariposa Formation (Upper Jurassic) one to two miles wide extends north through the central portion of the district. Greenstone and amphibolite are to the west, and schist and slate of the Calaveras Formation (Carboniferous to Permian) and granodiorite lie to the east. The Tertiary South Fork of the American River, which has numerous tributaries, entered the Placerville basin from Newtown. In places the Tertiary gravels are overlain by thick beds of rhyolite tuff and andesite.

Ore Deposits. Of the numerous tributaries of the main Tertiary channel in this district, probably the best known and one of the richest was the Deep Blue Lead. This channel extended south from White Rock to Smith's Flat and then west-southwest through the Texas Hill area. The lode-gold deposits are massive quartz veins as much as 20 feet thick with numerous parallel stringers. The ore bodies are low to moderate in grade (1/7 to 1/4 ounce of gold per ton), but the veins have been mined to depths of 2000 feet. The ore contains finely disseminated free gold and small amounts of pyrite. The veins occur chiefly in slate.

Mines. Placer: Coon Hollow $10 million, Diamond Springs, Green Mountain, Negro Hill,

Sacramento Hill, Spanish Hill $6 million, Smith's Flat $2 million+, Texas Hill, White Rock $5 million. Drift: Benfield, Cedar Spring, Clark, Kumfa, Landecker, Lyon, Pascoe, Rivera, Texas Hill, Try Again, Union.

Rattlesnake Bar

Location and History. Rattlesnake Bar is in northwestern El Dorado County and southern Placer County. The placer mines along the American River were highly productive during the gold rush. The town was established in 1849 and became goodsized until 1864, when it was destroyed by fire. The Zantgraff mine, the principal lode mine in the district, was active from 1880 to 1901 and again in the 1930s. Dragline dredging was done in the region during the 1930s. Part of the district is covered by the Folsom Reservoir.

Geology. The district is on the eastern flank of a major granodiorite stock that is intrusive into greenstones and amphibolite. A major body of serpentine and a limestone lens crop out in the area. Several extensive deposits of Pleistocene shore gravels along the American River were hydraulicked. The Zantgraff vein contains abundant sulfides, including galena and chalcopyrite, and was mined to a depth of 1100 feet. This district also has yielded substantial amounts of chromite and limestone and some copper.

Spanish Flat

Location and History. This well-known highgrade district is in west-central El Dorado County in the vicinity of the old mining town of Spanish Flat. It is about 10 miles north of Placerville and 3 miles northeast of Kelsey. The district was placer-mined soon after the beginning of the gold rush, and there has been intermittent mining ever since. Many of the early-day miners in this area, from South America, Mexico, and Portugal, were that time collectively known as "Spanish". The Alhambra mine, the largest source of gold in the district with a total output of $1.25 million, was active in the 1930s and 1940s and yielded much high-grade ore. Other mines in this district include the Brust, Shumway, and Timm mines.

Geology. The gold-bearing veins in this district are in a northwest-trending belt or zone about two miles east of the Mother Lode belt. The deposits consist of northwest-striking quartz veins with numerous parallel stringers that occur in shear zones with gouge. Country rock is amphibolite, chlorite, and graphite-quartz schist and slate. The ore shoots are not usually too extensive, but some have been extremely rich. A high-grade pocket discovered in the Alhambra mine in 1939 yielded $550,000. This pocket was a mass of native gold in quartz nearly 5 feet wide. The greatest depth of development is about 500 feet.

Volcanoville

Location. The Volcanoville district is in northcentral El Dorado County and south-central Placer County, about eight miles northeast of Georgetown and 30 miles cast of Auburn. It includes the Kentucky Flat area. It is both a lode- and placer-mining district.

Geology and Ore Deposits. A number of patches of andesite-capped Tertiary gravel were deposited by a channel that extended north through Kentucky Flat and then west through Volcanoville. An older well-cemented "white" channel contains much quartz gravel; there is a younger channel. The gold is coarse. Several gold-quartz veins in schist and state, near or adjacent to a belt of serpentine, crop out in the west portion of the district. Some of the quartz veins have yielded high-grade pockets and well developed quartz crystals.

Mines. Placer: Bedrock, Buckeye Point, Kenna, Kenny, Kentucky Flat, Morris, Tiedeman.

• Placer Mining Operations in El Dorado County

Big Canyon Dredging Company, Fresno, operated a dragline dredge on Deer Creek. During 11 months in 1941, operations yielded 3,160 ounces of gold and 321 ounces of silver from 540,000 cubic yards of gravel. During the first 7 months of 1942, operations yielded 1,269 ounces of gold and 137 ounces of silver from 250,000 cubic yards of gravel.

Duffy Property. George L. Duffy, Foresthill, holds placer mining claims in sec. 24, T. 13 N., R. 9 E., sec 3, T. 13 N., R. 11 E., M.D.M., and adjoining sections, to cover a dredging project on Middle Fork American River.

El Dorado Dredging Corporation, Greenwood, operated a dragline dredge on Greenwood Creek and on Coloma Crek during 1940. In 1941, operations between January 1 and March 6 on Coloma Creek yielded 833 ounces of gold and 124 ounces of silver from 106, 078 cubic yards of gravel. Equipment was moved to the Hughes property on Rock Canyon Creek, where 338, 940 cubic yards of gravel yielded 2,630 ounces of gold and 281 ounces of silver.

General Dredging Corporation, Natoma, operated a dragline dredge in 1940 and part of 1941 on the South Fork of American River near the point where James Marshall discovered gold in 1848.

Greenhorn Dredging Company, Youngs, operated a dragline dredge on the Middle Fork of Cosumnes River near Youngs during part of 1940 and 1941.

Hook and Ladder mine at Smiths Flat, sec. 10, T. 10 N., R. 11 E. Bert Bryan worked the Gray Lead channel on this property until 1932. An old shaft 114 feet deep was available at a point 350 feet south of the State highway. He sank an additional 38 feet, ran 330 feet south in bedrock, and raised to the channel, which he worked for a length of 1,150 feet. He states that gravel was 70 feet wide and 1-1/2 to 3-1/2 feet deep at this point. He thinks that several hundred feet of this channel remain in place, but that it is cut off by the Deep Blue Lead.

Patchen Property. Charles Patchen, Placerville, holds 80 acres in sec. 18, T. 10 N., R. 9 E., containing 1/4-mile of stream gravel 100 to 300 feet wide and 8 feet deep in Martell Ravine.

Starbuck Property. Frank M. Starbuck of Rescue holds 120 acres in sec. 16, T. 10 N., R. 9 E., which was worked by placer miners in the early days. Two to 6 feet of gravel are covered by 6 feet of overburden. Part of the tract near the mouth of Sweetwater Creek was worked recently by dragline dredge. Marcus Starbuck holds 40 acres in the adjoining section 17. Gravel is 250 feet wide and 10 to 15 feet deep. Bedrock is decomposed granite.

Two Channel Mine. This is a consolidation of several old mines. More than 2000 acres in secs. 9, 10, 15, 22, 27, 34, T. 13 N., R. 11 E., 8 miles northeast of Georgetown are included. A channel containing white quartz boulders, and another containing cemented gravel and andesite boulders are found here. Both have been worked by hydraulic and drift mining. Drifting has been done for a distance of 2500 feet south from Otter Creek.

Ventura Mine of 90 acres in sec. 20, T. 10 N., R. 12 E. A tunnel has been driven through a ridge containing a lava-capped channel.

Wulff Property. W. C. Wulff of Rescue holds 35 acres of placer ground in sec. 8, T. 10 N., R. 9 E. The productive part is 30 to 36 inches deep. The gold has evidently come from eroded pockets in the immediate vicinity. [1]

FRESNO COUNTY

Gold

• In the early days and up to as late as 1903 more or less placer mining as well as quartz, was carried on in many of the gulches and canyons in the foothill region of the Sierras in Fresno County. Among these may be mentioned particularly Temperance Flat, Sampson's Flat, Big Dry Creek district, and along the San Joaquin River around old Fort Miller above Friant (formerly Pollasky). Most of the placers, being only superficial, have been worked out.

Richard Burton and Joe Kesterman are working a placer claim (1915) on Laurel Creek in the NW 1/4, Sec. 11, T. 10 S., R. 26 E., 4 miles northeast of the Dinkey Creek Ranger Station.

Dodge Placer. John Dodge, owner. This placer mine at Kaiser Creek Diggings, in Sec. 4, T. 7 S., R 25 E., 12 miles due north of Cascada (Big Creek post office), has been worked in a small way by the owner for several years past. Elevation 5500 feet.

Robt. Ingelhardt is working a placer claim on Kaiser Creek north of Cascada.

Iowa, a placer property on the San Joaquin River about 6 miles above Friant. Idle past eight years (1915). It was worked by ground sluicing, using a turbine wheel and six 6-inch centrifugal pumps.

T. G. Price has been working a placer claim for several seasons on Dinkey Creek, 7 miles below Forest Ranger Station, about Sec. 3, T. 11 S., R. 26 E.

H. Richter is working a placer claim at Cabin Meadow, in Sec. 14, T. 10 S., R. 26 E., on Laurel Creek, northeast of the Dinkey Creek Forest Ranger Station.

William Wakefield is working a small placer property in Sec. 13, T. 10 S., R. 26 E., on a branch of Laurel Creek, 4 miles due east from the Dinkey Creek Forest Ranger Station. Elevation 7000 feet. [3]

Big Creek

Location. Big Creek is in eastern Fresno County about 10 miles northeast of the town of Trimmer and 50 miles northeast of Fresno. Superficial placer mining was done here during the early days and sporadic lode mining from the 1890s to around 1915.

Geology. A few narrow quartz veins contain free gold and often abundant sulfides. The country rock is granite and schist. A number of contact metamorphic zones in the region contain tungsten and some rare barium minerals, including sanbornite. The chief gold sources were the Contact and Hancock mines.

Big Dry Creek

Location. This district is in northeastern Fresno County about 10 miles northeast of Clovis and two miles north of Academy. The area was placer-mined on a small scale during the gold rush and lode-mined from the 1870s to around 1900. Apparently very little mining has been done here since.

Geology. The area is underlain by granitic rocks with thin bands of slate and mica schist. A serpentine belt lies to the east. A number of narrow north-trending quartz veins contain free gold and varying amounts of sulfides. The veins occur near or at the contacts between granitic and metamorphic rocks. The ore bodies are shallow, none having been mined to a depth greater than 150 feet. Copper also has been mined here.

Mines. Blue Rock, Confidence, Crystal Springs, Defiance, Monte Cristo, Rip Van Winkle, Thorn.

Friant

The Friant district is in northeastern Fresno County on the San Joaquin River, in the vicinity of the Friant Dam. Placer mining was carried on in the district during the early days in the vicinity of Fort Miller, a military post now inundated by the reservoir. Later this place was the terminus of a branch of the Southern Pacific Railroad and known as Pollasky. It was renamed in the 1920s for Thomas Friant of the White Friant Lumber Company. About $200,000 worth of gold was recovered as a by-product from sand and gravel excavated for use in the construction of the dam in 1940-42. Since 1946, from $5000 to $25,000 worth of gold has been produced annually from the sand and gravel plants here. The gold occurs in the river gravels and small terrace deposits adjacent to the river. The gold is fine and flaky, recovered in riffles set below the fine screens in the sand-washing plants.

Mill Creek

Location. This is a small district in southeastern Fresno County in the vicinity of the town of Dunlap about 40 miles east of Fresno. Superficial placer mining was done in Mill Creek and other streams during the early days. Small-scale lode mining was done from the 1880s through the 1900s, but little or nothing has been done since.

Geology. The area is underlain by granodiorite and related rocks with narrow slate belts. Some limestone lenses and schist are present. A few shallow north-trending veins a few feet thick contain free gold and varying amounts of sulfides. The veins usually are at or near granite-schist contacts. Most of the output has come from the Dixie Queen and White Cross mines.

Sampson Flat

Location. Sampson Flat is in eastern Fresno County about eight miles north of Dunlap and 45 miles cast of Fresno. The district includes the area known as Davis Flat. Sometimes it has been included in the Mill Creek District. It was placer-mined in the early days, and the lode mines were active front the 1880s until about 1915.

Geology. The district is underlain by granodiorite, gabbro and small amounts of pyroxenite. Several north-trending quartz veins two to three feet thick contain free gold and varying amounts of sulfides. There is some tungsten mineralization in bodies of tactite.

Mines. Davis Flat, Delilah (Black Jack & Hercules), Little Monitor, Oro Fino, Sampson.

Sycamore Flat

Location. Sycamore Flat is in east-central Fresno County just north of Piedra and about 25 miles due east of Fresno. It also is known as the Hughes Creek District. Superficial placer mining was done here during the gold rush, and the lode mines were active from the 1880s until about 1915. There has been minor prospecting since then.

Geology. The area is underlain by schist on the west, gabbro in the central portion, and granite in the east. There are a number of aplite dikes. Several narrow north-trending quartz veins with gentle to steep dips contain free gold and varying amounts of pyrite, chalcopyrite, and galena. A few high-grade pockets have been found here. One of the veins was mined to a depth of 300 feet.

Mines. Eliza Jane $100,000+, Independence, Sunnyside.

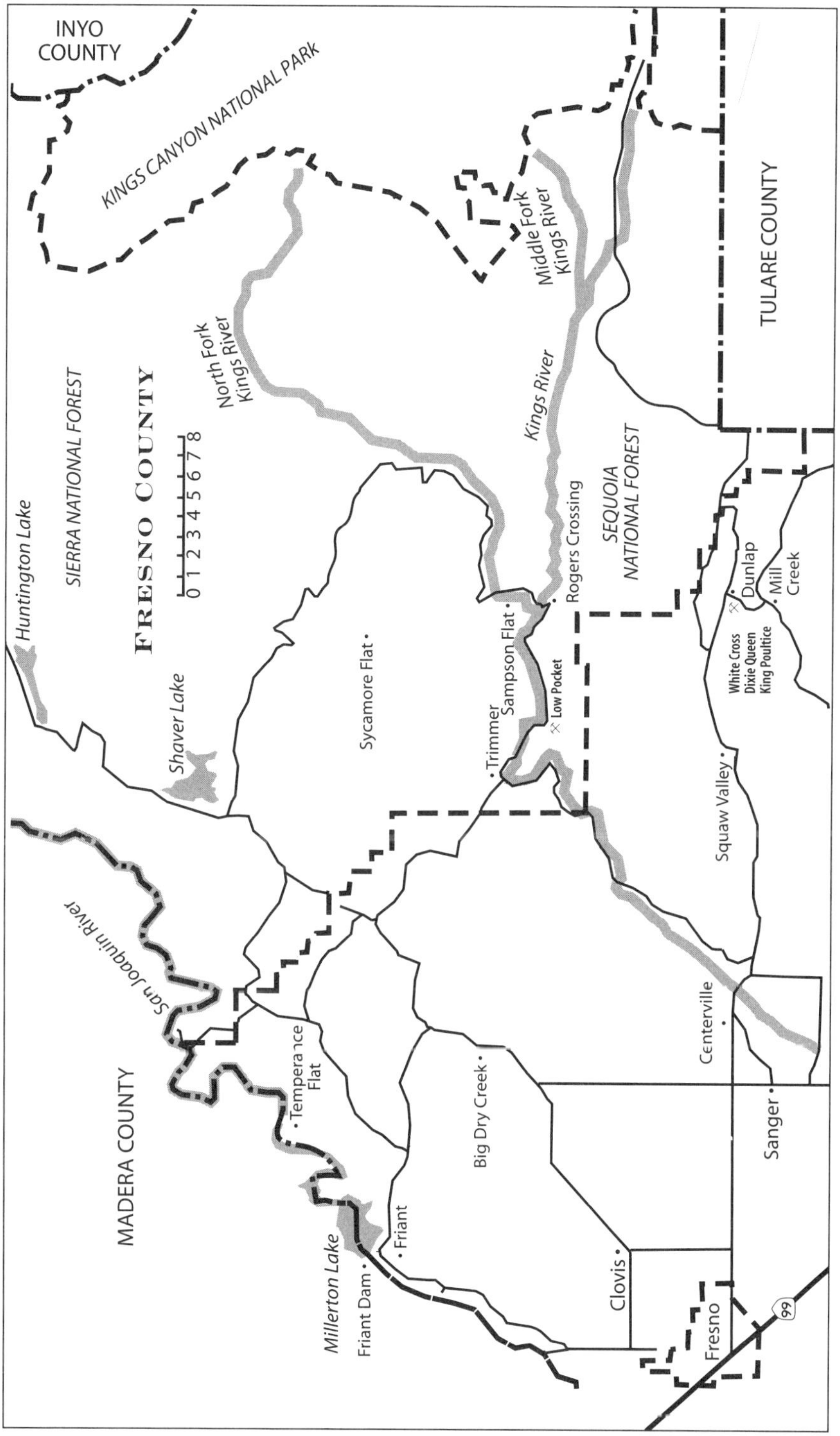
INYO COUNTY
KINGS CANYON NATIONAL PARK
SIERRA NATIONAL FOREST
Huntington Lake
FRESNO COUNTY
0 1 2 3 4 5 6 7 8
Shaver Lake
North Fork Kings River
Middle Fork Kings River
Kings River
TULARE COUNTY
Sycamore Flat
Sampson Flat
Trimmer
Low Pocket
Rogers Crossing
SEQUOIA NATIONAL FOREST
Dunlap
Mill Creek
White Cross
Dixie Queen
King Poultice
Squaw Valley
San Joaquin River
MADERA COUNTY
Temperance Flat
Big Dry Creek
Centerville
Sanger
Millerton Lake
Friant
Friant Dam
Clovis
Fresno
99

Temperance Flat

Location. Temperance Flat is in northeastern Fresno County on the south side of Millerton Lake. It is 10 miles northeast of Friant and about 25 miles northeast of Fresno. The area was placer-mined in the early days. Lode mining began at the Sullivan mine in 1853 and continued intermittently until about 1915. The area was prospected again during the 1930s.

Geology and Ore Deposits. The chief rock types are coarse-grained granite and granodiorite with diorite inclusions. Portions of the area are capped by thick flat beds of basalt of Table Mountain. A number of north-trending quartz veins, in shear zones in granitic rock, contain face gold and often abundant pyrite. Small amounts of other sulfides are present. A few small high-grade pockets containing leaf gold have been found here.

Mines. Henrietta, Keno, Quien Sabe, Providence, Rattlesnake, San Joaquin, Sullivan $100,000, Temperance, White Mule.

KERN COUNTY

• The first discovery of gold in Kern County was made by a member of Fremont's party in 1851, at Greenhorn Gulch, near the Kern River. Here a camp soon sprang up, and it is claimed that some of the placer ground yielded $50 to the pan. This territory was known in the early days as Whiskey Flat. [3]

Cove

Location and History. The Cove district is in northeastern Kern County between the towns of Kernville and Isabella on the west side of the Isabella Reservoir.

The upper Kern River here was placer-mined during the 1850s. The Big Blue vein was discovered by Lovely Rogers in 1860, and much activity followed during the 1870s and early 1880s. The mines were worked intermittently from then through the 1930s, the Big Blue group having been operated on large scale from 1934 until 1943. There has been only minor activity since. The main mines of the district have been consolidated into the Big Blue-Sumner group, which includes the Lady Belle group. The district has an estimated output valued at $8 million.

Geology and Ore Deposits. The principal mines are in granodiorite. To the east and south are schist, phyllite, quartzite, and marble of the pre-Cretaceous Kernville Series. Aplite dikes are often associated with the gold-bearing veins.

The ore deposits consist of extensive vein systems as much as 150 feet wide that occur in shear zones in granodiorite. The ore consists of quartz with finely disseminated free gold, arsenopyrite, pyrite, chalcopyrite, and galena. The mining ore usually averages 1/10 to 1/3 ounce of gold per ton with some highgrade streaks. Some scheelite also is present in the gold ore. The veins have been mined to depths of about 500 feet. There are two main vein systems: those of the Big Blue-Sumner and the Lady Belle groups.

• Willow (placer), consists of 40 acres in Sec. 22, T. 25 S., R. 33 E., M.D.M., in the Cove district, about 3/4 of a mile north of Kernville, at an elevation of 2700 feet. Owners, Orejana Mining Company, of Hayward. C. S. Long, president and manager. River deposit, slate and granite bedrock, soft. Pay gravel is about 6 feet deep. Extent of deposit not determined. Ditch has a capacity of 300 inches, the water being diverted from the Kern River. Worked several years ago by ground sluicing. Idle. [3]

Greenhorn Mountain

• Located about 48 miles north of Caliente, in T. 26 and 27 S., R. 31 and 32 E. A limited amount of placer mining is pursued in this district. The altitude ranges from 3000 to 6500 feet. [3]

Location and History. This district is about 28 miles northeast of Bakersfield. The first discovery of gold in Kern County was made in Greenhorn Creek in 1851 by a member of General John C. Frémont's party. A rush soon followed, and the town of Petersburg was established. Gold-mining activity declined before 1890, but there has been minor prospecting since. Most of the output has been from placer mining.

Geology. Much of the area is underlain by quartz diorite. There are a few bodies of metamorphic rocks and also some pegmatite dikes. The chief placer deposits were in Greenhorn, Fremont, Bradshaw, and Black Gulch Creeks. There are numerous small, poorly mineralized quartz veins, most of which are a few miles east of David Guard Station. The gold is in the free state and there is very little sulfide mineralization. Uranium-bearing peat bog was discovered in 1955 in the northwest part of the district.

Kern River

The upper Kern River between Bakersfield and Bodfish was the scene of a rush soon after the discovery of gold at Greenhorn Creek in 1851. However, the deposits in the river are believed to have been worked out in a short time. Many lode-gold prospects are in the area, but the only one of any consequence is the Gem gold mine near Democrat Springs. Uranium was discovered at Miracle Hot Springs in 1954, and there was a "boom" that lasted for a few years. Practically the entire region is underlain by quartz diorite.

Keyesville

Location and History. This district is in the southern Sierra Nevada in Kern County about 32 miles northeast of Bakersfield and two miles southwest of Isabella Dam. Gold was discovered here in 1852 by Richard M. Keyes, and for a time this was the largest community in Kern County. The chief periods of mining were the 1850s, 1860s, 1890s, and 1909-15. The area was prospected during the 1930s, but little has been done here since, and Keyesville has become a ghost town.

Geology and Ore Deposits. Virtually the entire district is underlain by quartz diorite. The gold deposits occur in a northeast-trending belt about three miles long. The veins consist of narrow quartz stringers with fault gouge that contain free gold and small amounts of pyrite, arsenopyrite, and pyrrhotite. There are some placer deposits, including one of possible Pleistocene age.

Mines. Bright Spot, High Grade, Homestake, Keyes $450,000, Keyesville, Keyesville Placer, Mammoth $500,000, Mooncastle, Nephi, Nob Hill, Opportunity, Sunrise, Virginia, Will Jean.

Keyes, the oldest mining district in Kern County, includes the territory north and west of the old site of Keyesville for a distance of about 5 miles, in T. 26 S., R. 32 and 33 E. Isabella, the post office, is 37 miles north of Caliente. Placer mining was extensively pursued here during the fifties.

Dead River Channel (placer), consists of 40 acres of placer ground in Sec. 17, T. 26 S., R. 33 E., M.D.M., in the Keyes mining district, about 1-1/2 miles north of Isabella, at an elevation of 2500 feet. Owner, C. E. Barton. Course of channel is north and south, depth of gravel 10 feet. Granite bedrock. Water obtained from Kern River. Small producer. Keysville Placer, consists of 40 acres in Sec. 36, T. 26 S., R. 32 E., M.D.M., about 3 miles southeast of Isabella, in the Sequoia Forest Reserve, owned by G. Henschkel. Elevation 3400 feet. Old river deposit, worked in winter by ground sluicing. Course of channel northwest and southeast, granite bedrock, pay gravel 5 to 10 feet deep, easily worked, small producer.

Long Tom

Location and History. The Long Tom district is in the southern Sierra Nevada in central Kern County. It is 23 miles northeast of Bakersfield and 10 miles south of Woody. The veins at the Long Tom mine, the chief source of gold in the district, were discovered prior to 1860 by prospectors looking for the source of placer gold in nearby creeks. The mine was considerably active during the 1880s and again from 1925 to 1939. It has an estimated total output of $800,000 to $900,000.

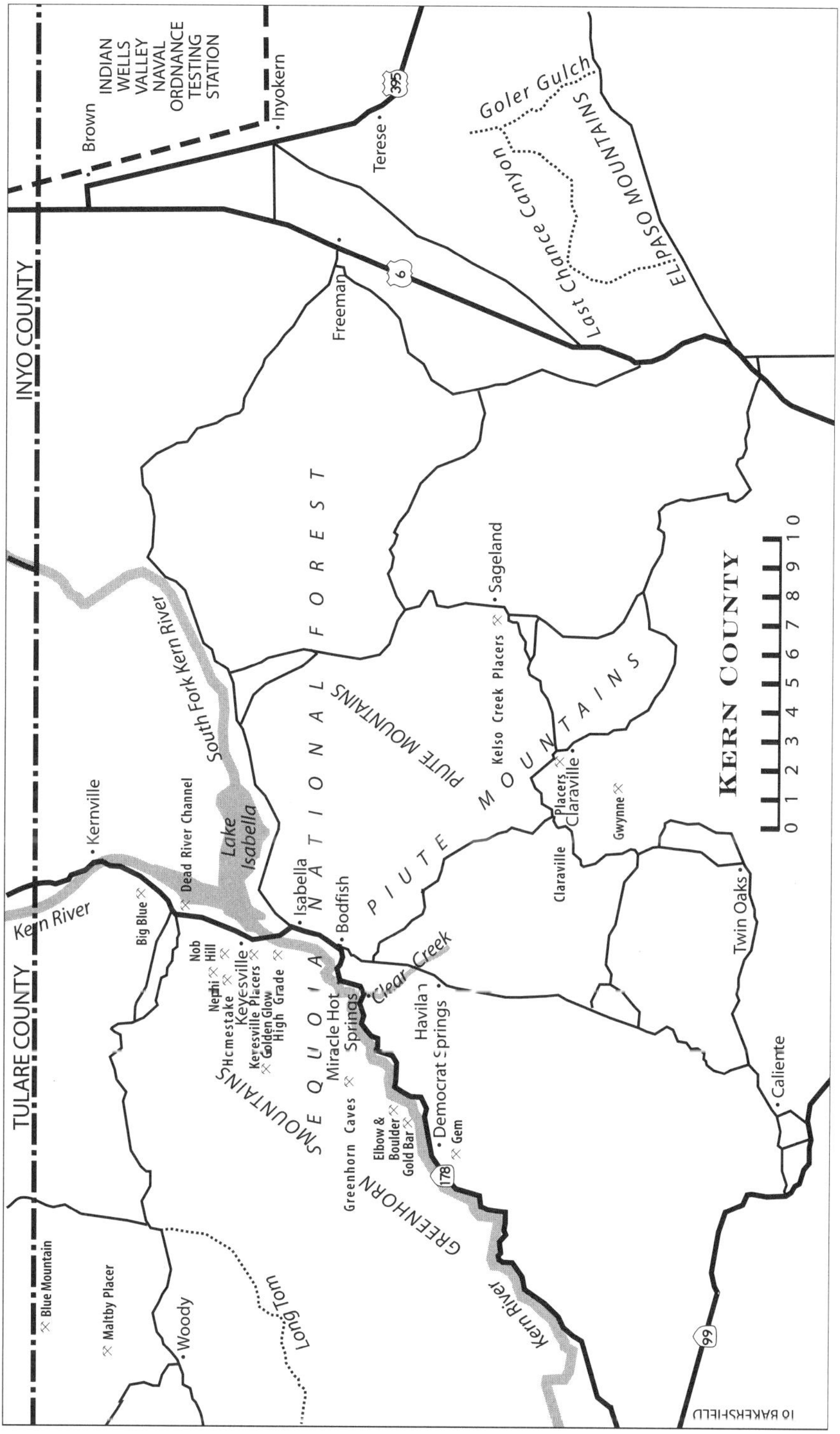
INYO COUNTY
TULARE COUNTY
INDIAN WELLS VALLEY NAVAL ORDNANCE TESTING STATION
Brown
Inyokern
395
Terese
Goler Gulch
Last Chance Canyon
EL PASO MOUNTAINS
Freeman
6
SEQUOIA NATIONAL FOREST
Sageland
Kelso Creek Placers
PIUTE MOUNTAINS
Placers
Claraville
Gwynne
Twin Oaks
Caliente
KERN COUNTY
0 1 2 3 4 5 6 7 8 9 10
South Fork Kern River
Kernville
Dead River Channel
Lake Isabella
Isabella
Bodfish
Kern River
Big Blue
Nob Hill
Nephi
Keyesville
Homestake
Keyesville Placers
Golden Glow
High Grade
Clear Creek
Havilah
Miracle Hot Springs
Democrat Springs
Greenhorn Caves
Elbow & Boulder
Gold Bar
Gem
178
GREENHORN MOUNTAINS
Blue Mountain
Maltby Placer
Woody
Long Tom
99
to BAKERSFIELD

• Clear Creek District

The Clear Creek or Havilah district is in northeastern Kern County about 26 miles east-northeast of Bakersfield and 5 miles south of Bodfish. It comprises an area of about 40 square miles which is bounded approximately by Hobo Ridge on the west, Kern River on the north, Walker Basin on the south, and Bald Eagle Peak on the east. The district has yielded principally gold and tungsten, but deposits of placer gold, antimony, uranium, and rare earths also have been found.

The earliest discoveries of gold at Clear Creek were made in 1864 by Claude de la Borde, George McKay, Benjamin T. Mitchell, and Hugh McKeadney (Boyd 1952, P. 39). In 1865 the town of Havilah was established at the center of an elongate north-trending valley which traverses the area, and by the winter of 1865-66 had attracted an estimated 3,000 people. In 1867 it became the first seat of Kern County, an honor which it held until 1874 when the county seat was moved to Bakersfield.

The Clear Creek district is underlain mostly by plutonic igneous rocks of Mesozoic age, which range in composition from quartz diorite to gabbro, although quartz diorite predominates. Roof pendants of pre-Cretaceous metasedimentary rocks lie along a narrow north-trending belt that traverses the north-central part of the district northwest of Havilah. The Red Mountain area, just north of Walker Basin, is underlain by similar rocks. These metasedimentary rocks strike generally north and nearly everywhere dip vertically. Several elongate, north-trending pegmatite dikes intrude the plutonic rocks east of Havilah Creek, which flows north through Havilah.

The mineral deposits, almost without exception, lie along north- to northeast-trending structural features. Most of the gold deposits are in a group of veins that strike N. 45° E. and dip steeply to the southeast. Although the veins are discontinuous, they form a zone about 1,000 to 2,000 feet wide that extends from the area just north of Flying Dutchman spring to the base of Rankin Peak, a distance of approximately 4 miles. Individual veins range in width from 3 inches to 6 feet, and most of them can be traced for several hundred feet along the surface. They are composed typically of quartz and fault gouge in which are small grains of arsenopyrite, pyrite, and free gold. Silver is generally present in small proportions, probably in solid solution with the gold, as no silver minerals have been recognized. Many of the ore shoots are at or near junctions of the principal vein with converging cross fractures, although some of the ore shoots have no apparent structural control.

Tungsten minerals are present in both quartz veins and tactite zones. The tactite bodies typically are lenticular and 1 to 15 feet wide. They are irregularly disposed along the margins of roof pendants, or are adjacent to small masses of granitic rock within the metamorphic rocks. In these bodies the tungsten is in disseminated grains of scheelite in a gangue composed principally of quartz, garnet, calcite, and epidote. The tactite can be recognized easily by its characteristic brown and green color. In the Clear Creek district tactite bodies formed as a replacement of limestone are more common than those formed by replacement of other sedimentary rocks.

The scheelite-bearing quartz veins of the district (Tungsten King, Bald Mountain group), although not in contact with the roof pendants, are within a few hundred feet of them. These veins, which are all in the vicinity of Hooper Hill (formerly Bald Mountain), strike about N. 30° E. and dip 60° to 80° SE. The veins contain disseminated scheelite crystals, and scattered crystals of pyrite.

Three-quarters of a mile northeast of Havilah at the Alice mine antimony is found in scattered patches within a north-trending pegmatite dike.

Radioactivity has been noted in several pegmatite dikes in the district and has been attributed, in at least one occurrence, to the presence of cyrtolite, a uranium- and rare-earth-bearing zircon. No economic concentrations of radioactive minerals have been found in these dikes.

Cove Mining District

by *Thomas E. Gay, Jr.*

The Cove district includes 30 or more claims on the west side of the Kern River Valley, 2 to 4 miles south of the new town of Kernville, near the west shore of Lake Isabella. Most of the more productive mines are gold mines in section 28, T. 25 S., R. 33 E., M.D.M. The Cove district has been referred to in recent reports as the Big Blue group because the Big Blue mine was one of the more important mines in the later active years of the district. A comprehensive report on the history, geology, and operations of the Big Blue group was written by J. W. Prout (1940). The dozen or more productive mines in the district were active during various periods from 1860 to 1943. The Big Blue mine was closed in 1943 by War Production Order L208 and no gold has been mined since then. The following description of the district is based on Division of Mines reports, mainly Dr. Prout's (1940), on property visits, and on conversations with Dr. Prout in 1955.

In 1851 the discovery of placer gold brought a rush of prospectors to the Kern River basin. The placer deposits proved to be disappointing in number and quality to most of the gold seekers, although daily recoveries of $16 to $60 were reported by a few miners. Placer values in the river banks and bars in the Cove district were low, and by 1855 most of the prospectors had departed.

In 1860, quartz-bearing float was traced to an outcrop, which became the discovery point for the Big Blue mine. The lusty mining camp of Quartzburg was established near the present site of the Big Blue mill, and the Cove district commenced a period of essentially continuous though incompletely documented activity, which lasted until 1883 when a serious mine fire occurred. Arrastras, Chilean mills, and, later, stamp mills with amalgamation machinery, powered mainly by Kern River water, were used to recover an estimated total of several million dollars in gold prior to 1883.

Details of the ownership and operation of the mines in the district prior to 1875 is largely undetermined. Lovely Rogers, who discovered the Big Blue vein, evidently did the first lode mining, about 1862. A score or more of claims along and adjacent to the Big Blue vein were staked by miners who worked them individually during most of the 1860s, and apparently made large recoveries from high-grade, near-surface ore bodies. Values were reported to be as high as $300 per ton in free gold. In 1868 Judge J. W. Sumner consolidated part of the mining interests and produced a substantial amount of gold. During the period 1856-76 the principal mine was reported to be the Sumner. The lode was 80 feet wide and average yield of the mined material was about $18 per ton (Raymond, 1875). An "old" 16-stamp custom (?) mill was said to have yielded a profit of $100,000 to $300,000.

In 1875-76 Senator John P. Jones of Nevada consolidated all water and mining rights in the Cove district and formed the Sumner Gold and Silver Mining Company. This company constructed and operated new 16- and 80-stamp mills, powered by a 54-inch water turbine. A steam plant was erected to power walking-beam Cornish pumps to dewater the six-compartment Sumner (or "Engine") shaft, which was then 400 feet deep and the principal mine opening.

In 1881 the holdings were leased to Michaels, Freidlander and Company, who operated them until late 1883 when the property was ravaged by fire. The underground workings continued to burn for several years and most of the workings above the 260 level caved.

From 1883 to 1934 the district was essentially unproductive, although several attempts were made to restore the mine to its former productivity. Between 1894 and 1900, probably about $50,000 in gold was recovered during intermittent operations. In 1924 the total production of the Cove district was estimated at $5,000,000 to $8,000,000 (Tucker, 1924, p. 35). From 1907 to 1926 the property was owned by the Kern Development Company, who reopened some of the lower workings and leased various portions of the property, but no commercial production resulted.

In 1927, a Mr. Jubien leased four of the previously more productive claims, but drifting and diamond drilling on the 260 level failed to reveal commercial ore bodies and the lease was relinquished.

From October 1931 to September 1932 the American Smelting and Refining Company leased the same four claims and explored by means of drifts and winzes on the 260 and 400 levels; again no production resulted.

In 1934 a group later known as the Big Blue Mining Company was organized by A. V. Udell to operate the property (Prout, 1940, p. 385). Another mill was constructed and considerable development workings were driven but the company became bankrupt and their interest in the properties was bought out in April 1935 by Kern Mines, Inc., who are the present owners (1958). From 1936 to 1943 deposits that center about the Big Blue claim were continuously worked, yielding several hundred thousand tons of ore from which gold valued at more than a million dollars, and lesser amounts of silver, lead, and copper were recovered. In 1943, the Government closing order L208 caused the mine to be shut down permanently.

The Big Blue mill and the portal of the mine drainage tunnel are at lower elevation than the spillway of Lake Isabella dam, which was constructed by the U.S. Army Corps of Engineers in 1948-53. Hence the workings would be subject to flooding when the reservoir is filled to capacity. In 1954, to complete the reservoir project, the Corps of Engineers acquired all land below elevation 2617, including the mill and drainage tunnel portal. In 1955 a watertight bulkhead was constructed across the drainage tunnel by the Corps of Engineers, and a pump installed to dispose of mine drainage waters. In 1957 the mill was sold at auction, and removed to New Mexico.

The principal rock units of the Cove district are the Mesozoic Isabella granodiorite, including aplitic and alaskitic phases, the Carboniferous (?) schist, quartzite , and phyllite of the Kernville series (Miller and Webb, 1940; Prout, 1940). Gold mineralization has been localized mainly within two intersecting systems of faults and shear zones. The principal vein system, the Big Blue or Sumner lode, is a sheared and faulted zone, in places as wide as 125 feet, that strikes about N. 30° E., and dips about 70° NW. This zone, known as the Big Blue-Sumner shear zone, can be traced laterally on the surface for more than 8,000 feet, and contains many subordinate faults, splits, and sheared zones. Another group of veins, the Lady Belle group, lies west of the Big Blue-Sumner shear zone. It dips 60° to 80° SW., and terminates against the west side of the northern portion of the Big Blue-Sumner shear zone.

Gold, ranging from 650 to 700 fine, occurs in both vein systems. In some places it is not associated with other metallic minerals. In other places it is accompanied by arsenopyrite, galena, and sphalerite. Other metallic minerals reported in the ore include magnetite, marcasite, pyrrhotite, scheelite, molybdenite, bismuthinite, sphalerite, and stibnite (Prout, 1940, P. 411-413).

The main mines of the Cove district can be divided into two groups: those on the Big Blue-Sumner shear zone, including, from south to north, Nellie Dent, Content, Blue Gouge No. 2, Big Blue, Sumner, and North Extension Sumner; and those on the Lady Belle vein system, including, again from south to north, the Lady Belle, Jeff Davis, Bull Run, Frank, Urbana, and Beauregard. Other claims adjacent to those listed above were worked, but few data are available concerning them. The available records are not entirely clear as to which portions of the district were active at particular times, nor as to which portions of the various ore bodies were the most productive.

The mines of the Lady Belle group of veins yielded high-grade ore from about 1862 through the early 1880s, but they have been practically inactive since then. The mines on the Big Blue-Sumner shear zone had about the same early history as those of the other group, but were also intermittently active in the early 1900s, with peak productivity from 1936 to 1943. Mines of both groups were worked individually in the early years of the district, but most of the production resulted at times when the district was consolidated under unified management, such as that of the Sumner Gold and Silver Mining Company (1875-81), and Kern Mines Incorporated (1936-43). Although the veins of the Lady Belle system differ geologically from the Big Blue-Sumner veins, the two systems were mined through the same workings after the fire of 1883 made most near-surface portions of the Lady Belle system inaccessible. Published records, which do not include details of the district's earliest years, suggest that the Big Blue-Sumner mine was the most productive. It has the longest history of activity, the greatest recorded production, the deepest and longest workings, and has been the main means of underground access to several adjacent claims during consolidated operations.

El Paso Mountains District

El Paso Mountains, which trend east-northeastward on the north edge of the Mojave Desert in northeastern Kern County, are about 25 miles long and 4 to 10 miles wide. Placer gold, pumice, pumicite, and clay are the principal mineral products of the area. Lode gold ore, coal, and copper ore also have been mined. Manganese, uranium, perlite, molybdenum, and lead prospects in the area have been explored. Some of the rocks may eventually be quarried as ornamental stone. The northwestern part of El Paso Mountains is a popular area for collecting fluorescent and fire opal, silicified palm, jasper, agate, and other material suitable for cutting and polishing. The gross value of all mineral products obtained from deposits in El Paso Mountains probably exceeds several hundred thousand dollars—mostly for placer gold, pumice, and pumicite.

The geology of El Paso Mountains has been mapped and described by Dibblee (1952). Metamorphosed sedimentary Permian (in part ?) rocks overlie Precambrian? schist and crop out in the northeastern part of the range. Quartzite conglomerate and hornfels (Jurassic-Triassic ?) crop out at the southwestern part of the range. These Precambrian? to Jurassic? rocks are intruded by Mesozoic rocks of quartz diorite to granite composition which crop out mostly in the southwestern part of the range. The Goler formation, of Paleocene age (McKenna, 1955, p. 512-515), unconformably overlies the older rocks and is in turn unconformably overlain by the Plio-Pleistocene Ricardo formation. Both of these Tertiary formations are composed largely of continental deposits and lie northwest of the main part of the range.

The metasedimentary and intrusive rocks are the host rocks for copper, gold, lead, manganese, and molybdenum mineralization. Sedimentary deposits of clay, coal, perlite, pumice, and pumicite are in the Ricardo formation, and uranium mineralization is associated with some strata in this formation. Placer gold is sparsely distributed in basal members of the Goler and Ricardo formations, but the Quaternary gravels are richer.

The placer gold deposits were discovered about 1893 and hundreds of prospectors obtained small amounts of gold by "dry washing" the gravels, mostly during the 1890s and a several-year period in the 1930s. Most of the clay was mined between 1920 and 1940 and was used principally as an adsorbent in refining petroleum products. Pumice and pumicite, mined nearly continuously since the 1920s, have been used in cleansing compounds and filler material. In 1958, the pumice deposit mined by Calsilco Corp. was the only active mine in the area.

The principal placer gold areas are, from east to west: Goler Canyon and Reed and Benson Gulches on the west side of Goler Canyon; the segment of Last Chance Canyon between Holloway Camp and the mouth of Bonanza Gulch; Bonanza Gulch; and tributary canyons on the east side of Redrock Canyon. The placer gold generally is most abundant in the lowermost part of Quaternary gravels on the sides of the stream beds or gulches. These gravels, or fanglomerate, contribute gold to smaller, but commonly rich, concentrations of placer gold accumulated in gullies downslope from the gravels.

Copper, gold, lead, manganese, and molybdenum mineralization is associated mostly with quartz veins in metasedimentary rocks and igneous rocks. Copper deposits are the most abundant of these and commonly contain traces of gold and silver. Lead is found in only a few prospects, and manganese was noted only in traces in some of the quartz veins. Molybdenum is present as powellite pseudomorphs after molybdenite and as molybdenite disseminated in dikes. Fragments of the powellite pseudomorphs have been collected for many years from gravels in upper Last Chance Canyon and are prized as mineral specimens.

The clay and pumice deposits in the lower part of the Ricardo formation are in a northeast-trending belt extending between Redrock and Black Mountain. In the southern part of the belt a layer of altered pumice and pumicite was the source of a few tens of thousands of tons of adsorbent clay obtained mostly from the Snow White mine. This same sedimentary layer, some 7 miles to the northeast, is the source of high-grade pumice which is marketed by Calsilco Corporation for several uses. A layer of very pure pumicite which occupies a stratigraphic position a few hundred feet higher in the Ricardo formation than the pumice layer was mined for nearly 25 years by Cudahy Packing Co. for use in a household cleaner.

Erskine Creek District

Erskine Creek, a northwest-flowing tributary to the Kern River, cuts through a north-trending roof pendant of pre-Cretaceous metasedimentary rocks that contains deposits of tungsten, gold, silver, antimony, uranium, copper, and building stone. These deposits lie within an area about 5 miles long and about 2 miles wide. Center of the area is about 5 miles southeast of Bodfish.

Antimony and gold deposits were productive in the early 1890s and later, though production was probably not large. Copper was discovered before 1904 but as late as 1958 the prospects remained unproductive. Tungsten was produced in the 1940s and 1950s, and probably has the highest total dollar value of materials mined in the district. Uranium mineralization was discovered in 1954, but only several tens of feet of underground workings were driven. Fine-grained metasedimentary rocks are mined intermittently from a deposit on Cook Peak and marketed as building stone. A travertine deposit at the mouth of Erskine Creek has been prospected for stone but none had been produced, by 1959.

Gold, associated with unidentified silver minerals, has been mined mostly from the Iconoclast mine from quartz veins in a northeast-trending zone that is nearly vertical.

The principal source of tungsten in the district is a deposit of scheelite in metamorphic rocks at the Unip mine. Scheelite is also in the tactite at the Christmas Tree prospect.

Antimony is found in a quartz vein at the Tom Moore mine in fine-grained metasedimentary rocks. The principal minerals are stibnite, native antimony, and several oxides of antimony.

Greenhorn Mountain District

The Greenhorn Mountain district comprises about 20 square miles of the Sierra Nevada approximately 28 miles northeast of Bakersfield and 4 miles northwest of Miracle Hot Springs. Its approximate boundaries are the Kern River on the south, Fremont Creek on the west, Woodward Peak on the north, and Black Gulch on the east. The first gold discovery in Kern County was made in Greenhorn Creek in 1851, by a member of General John C. Fremont's exploration party. An influx of prospectors followed, the town of Petersburg was established near Fremont Creek, and the district was extensively developed. After a disappointing recovery of placer gold, however, interest declined, and by 1900 Petersburg had become a ghost town. Subsequent intermittent and small-scale attempts to mine gold from lode and placer deposits have been mostly unsuccessful. As most of the mining in this district was done prior to 1890, no accurate production statistics are available.

Mesozoic igneous rocks, largely quartz diorite, underlie most of the area, although small roof pendants of pre-Cretaceous metasedimentary rocks and elongate north-trending pegmatite dikes crop out at scattered localities.

Only lode and placer gold have been mined in the Greenhorn Mountain district, although an unusual uraniferous peat bog deposit was discovered in 1955 in the extreme northwestern part of the area. The lode gold deposits are in northeast-striking fissure veins which range in width from a few inches to several feet. The veins strike N. 35°-75° E. and dip steeply northwest or southeast. They consist typically of iron-stained fault gouge and minor amounts of quartz. Free gold is the only ore mineral. Wall rocks generally consist of decomposed quartz diorite, but in a few mines one wall is composed of pegmatite.

Most of the veins are within a 4-square-mile area, the center of which is the 2 miles due east of David Guard Station. The veins are quite numerous and widespread but generally are weakly mineralized, narrow, and discontinuous. These features are reflected by the many shallow prospect holes and in the low productivity of the district.

Placer deposits derived from a large volume of bedrock probably were the source of most of the gold mined in the district. The discovery of a few rich placers apparently provided impetus to mostly unsuccessful attempts to find equally rich vein deposits from which the gold was derived.

The uraniferous peat bog deposit is on the old Pettit Ranch at the head of Little Poso Creek. Unidentified uraniferous material is still being deposited from uranium-bearing spring water into a bog consisting of woody fragments, black carbonaceous matter, silt, and decomposed granitic fragments. No mining of the material had been attempted by the end of 1958.

Kern River Canyon District

The Kern River Canyon area discussed herein is a 20-mile strip from 1 to 2 miles wide, extending from Bodfish southwestward along the Kern River to the west front of the Sierra Nevada, about 10 miles northeast of Bakersfield.

Placer gold and uranium deposits, mostly in the northeastern half of the area, have been mined, and lode gold as well as lode and placer tungsten deposits have been explored. Placer gold deposits were mined as early as 1851 in Greenhorn Creek (Gulch) near its confluence with the Kern River (Tucker and Sampson, 1933, p. 278). Although little is known of the area's early placer-mining activities, the deposits apparently were few and soon worked out. The most productive placer gold mine apparently was the Greenhorn Caves mine in Greenhorn Creek with a reported production valued at $60,000 (Tucker and Sampson, 1933, p. 307). No reliable estimates of total production from placer mines in the area is available. Of the numerous prospects for lode gold and tungsten, only the Gem mine was of any consequence. At the Gem mine, 1 mile southwest of Democrat Springs, reported production was valued at $30,000 (Tucker and Sampson-1933, P. 303).

The most recent mining interest in the Kern River Canyon area is in the uranium deposits, first noticed by Harry B. Mann in January 1954 (Bowes, 1957, p. 7). Through 1958, three deposits, the Kergon, Little Sparkler, and Miracle mines, have yielded a total of about 11 railroad cars of ore which averaged about 0.3 percent U3O8.

Almost the entire Kern River Canyon area is underlain by Mesozoic quartz diorite and associated aplitic and pegmatitic dikes, which most commonly trend north. Small pendants of pre-Cretaceous metasedimentary rocks are present in the granitic rocks. Moderate foliation is common in the granitic rocks, especially near contacts with the pendants.

The gold and tungsten placer deposits are in the recent gravels of the Kern River and its tributaries. The gold is probably derived from the Greenhorn Mountain, Keysville, Clear Creek, and the Cove districts, which lie within a 10-mile radius to the north and northeast. Most of the scheelite probably is derived from the Clear Creek district, although Erskine Creek and Kernville areas may have contributed substantial proportions. No placer tungsten has been shipped.

Uranium mineralization apparently is confined to a 1-square-mile area centered about 1-1/2 miles west of Miracle Hot Springs. The uranium is found along fractures and shears which have no apparent favored orientation. Ore deposition was controlled by intersection of shears and fractures or by abrupt changes in the dip of the vein (Bowes, 1957, p. 5).

Keysville District

The Keysville district is 32 miles northeast of Bakersfield and 2 miles southwest of Isabella Dam. It comprises about 15 square miles, bounded approximately by the Kern River, French Gulch, and Black Gulch.

Gold was first discovered in this district in 1852 by Richard M. Keyes (Tucker and Sampson 1933, p. 278) who located a group of claims later known as the Keyes mine. This and the Mammoth mine, which was located soon thereafter, became the two most productive mines of the district. Subsequent discoveries at the sites of the Pennsylvania, Sunrise, and other mines led to the establishment of Keysville, the ascendant community of the Kern River country until the boom at Havilah in 1867. Keysville, which is 2 miles southwest of Isabella Dam on the west side of Kern River, has since become a ghost town. The principal periods of mining activity after the 1860s were during the 1890s, from 1909 to 1915, and for a brief time following the rise in the price of gold in 1933.

Mesozoic plutonic igneous rocks, predominantly quartz diorite, underlie virtually all of the area. The gold is in a vein system which strikes N. 40°-50° E. and dips 70°-80° SE., closely resembling in nature and attitude the vein system in the Clear Creek district 7 miles to the south. Both systems contain sub-parallel discontinuous quartz veins, and both apparently terminate at the Kern Canyon fault. The Keysville veins, unlike those at Clear Creek, contain little pyrite and arsenopyrite.

Most of the deposits lie in a mile-wide northeast-trending belt extending 3 miles southwest from Lake Isabella to a northwest-trending ridge overlooking Black Gulch. The veins range in width from 3 inches to 6 feet and each consists typically of a gouge-filled fault zone which contains narrow, discontinuous quartz stringers. Gold is in the free state, both in the gouge and in the quartz stringers. In some veins it is associated with sparsely distributed grains of pyrite, arsenopyrite, and pyrrhotite. Hydrous iron oxide stains are common.

Locally derived Recent placer deposits, which are mostly in shallow draws and intermittent streams, have been mined for gold since 1852. One mine, however—the Keyesville Placer—is in Pleistocene? auriferous stream gravels which are poorly sorted and occupy a southeast-draining channel about a mile southeast of Keysville. This channel lies above the present channels and has been dissected almost to bedrock. The gravel deposit is apparently 10 to 20 feet thick, less than 20 feet wide, and less than 2,000 feet long. Its known production is small.

Piute Mountains District

The Piute Mountains district, an area of about 35 square miles, contains about 40 mines and prospects. The district is in the crestal part of the Piute Mountains and centers about the townsite of Claraville, about 14 miles southeast of Bodfish. Gold and tungsten have been the principal mineral products; antimony was produced from one deposit; and large bodies of white to gray carbonate rock have been prospected.

The most productive mines in the district have been the Bright Star, which was discovered about 1870, and yielded gold valued at about $600,000 (Brown, 1916, p. 490) and the Gwynne, which yielded about $770,000 in gold (Tucker, Sampson, and Oakeshott, 1949, p. 244). Each of the two has yielded several times more gold than the combined total of the other gold mines in the district. The principal periods of gold mining were 1870 to 1900, and the 1930s to the 1940s. An undetermined but small tonnage of tungsten has been produced—probably from three mines—since about 1950, and antimony valued at $13,000 was mined from the Jenette-Grant mine in 1918 (Goodwin, 1957, p. 529).

Most of the district is underlain by Mesozoic granitic rocks which contain a pendant of pre-Cretaceous metasedimentary rock. The pendant is from half a mile to 2 miles wide and is several miles long.

The gold deposits are in a belt about 2 miles wide that trends northwest in the granitic rocks, and north-northwest in the metamorphic rocks. Most of the gold deposits are in quartz veins which occupy shear zones and closely spaced joints, and trend mostly northeast. In most mines the gold is free-milling. Sulfide mineralization is evident in the deeper parts of some of the mines. According to U.S. Bureau of Mines production records, ore from most of the mines contained from 1/6 oz. to 11-1/2 oz. of gold per ton, which accompanied silver at a ratio of about 2:1. About 1/2 oz. of gold per ton is probably the approximate average for all mines.

The tungsten deposits are of two types: scheelite in gold-bearing quartz veins, and scheelite in tactite bodies near limestone. Most of the tungsten deposits explored by mid-1959 yielded pods of ore rarely more than a few feet in maximum dimension.

Antimony was mined along a north-trending contact between limestone and phyllite at the Jenette-Grant mine. The principal mineral was stibnite.

Carbonate rocks, in masses as much as several hundred feet long, several tens of feet wide, and several tens of feet in exposed vertical extent are found in the roof pendant of metasedimentary rocks. A few small open cuts have been made in the whitest portions of some of the bodies along the east side of the pendant. Rocks exposed in these cuts are white enough to be utilized as roofing-granule material, and some might be suitable for filler material.

Woody (and White River) District

The Woody district is in northern Kern County about 30 miles northeast of Bakersfield. The area embraces about 70 square miles included in T. 25 and 26 S. and R. 29 and 30 E., M.D.M. Most of the mines are clustered in two areas—one just south of Woody, the other 2 to 3 miles south of White River near the Tulare County line. In the following text they are considered as one district.

Deposits of gold, copper, tungsten, and iron are known to exist in the area, but production records and information concerning the early history of mining are lacking. Gold was mined in the district as early as the late 1850s, and most of the total gold output of the district probably was produced before 1890. Only the Blue Mountain mine has been worked seriously since 1890. It was active during the period 1909-23 and again briefly from 1929-31 (see tabulated list). Tungsten mines in the district were most active during the periods 1940-43, and 1952-56, when more than 15,000 units Of WO3 was produced. The Greenback Copper mine, a quarter of a mile south of Woody, producer of almost all the recorded copper output from the district, yielded an estimated 600,000 pounds of copper during the years 1890 to 1900 and 1913-18. The only body of iron ore in the district is at the Iron Mountain deposit where concentrations of magnetite in schist have been reported. No production has resulted from the limited exploration done at this property.

Granitic rocks of the southern Sierra Nevada batholith underlie most of the Woody district, but a few roof pendants composed of metasedimentary rocks form some of the more resistant peaks. These pendants range from a few tens to a few hundreds of feet in width and from a few hundreds to a few thousands of feet in length. Oligocene nonmarine sedimentary rocks border the district on the west, but no ore deposits are known to be associated with these rocks.

Gold

In terms of total dollar value and number of deposits, gold is the most important metallic mineral commodity that has been produced in Kern County. From 1851, when it was discovered in Greenhorn Gulch near the Kern River, through 1957, the value of gold mined in Kern County exceeded $46,000,000.

The first lode mining was in 1852 at the Keyes and Mammoth mines at Keysville, and by 1865 gold was being produced from at least four districts in the Kern River country: the Keysville, Clear Creek (Havilah), Greenhorn Mountain, and Cove districts. Gold was so important to the economy of Kern County during this period that Havilah, a remote settlement 7 miles south of Lake Isabella, was the county seat from 1867 to 1874. The gold and silver produced from these districts has been estimated at several million dollars, although no production records are available for the period before 1880. In 1894, gold was discovered on Standard Hill in the Mojave district; through 1958, gold and silver valued at about $20,000,000 was produced from the four isolated buttes that comprise the district. Discovery of gold at the site of the Yellow Aster mine in 1895 led to development of the Rand district and an eventual total yield of at least $20,000,000 in gold and silver.

Two mines, the Yellow Aster mine ($12,000,000 output) in the Rand district and the Golden Queen Mine ($9,000,000 output) in the Mojave district have yielded almost half of the total recorded gold output of the county.

Most of the gold veins in the Rand district occupy shear zones and faults in Rand schist (Precambrian?) and Atolia quartz monzonite (Mesozoic). In the Yellow Aster mine, the gold also is in a network or stockwork of closely spaced veinlets, which were mined almost as if they formed a single large vein.

In the Mojave district gold is in veins in Tertiary rhyolitic volcanic rocks, although in some deposits quartz monzonite is the host rock on one or both walls. Associated with gold in the veins are silver, copper, antimony, and lead minerals.

Although most of the gold deposits in the county are in the Sierra Nevada, these deposits have a lower average yield than those in the Mojave Desert region and they are more widely spaced. The nine principal districts in the Sierra Nevada have a combined minimum output of about $7,000,000 in gold. These districts in the approximate order of decreasing productivity according to recorded production are: the Cove, Keysville, Clear Creek (Havilah), Loraine, Pioneer, Piute Mountains (Green Mountain), Poso Creek, Woody, and Greenhorn Mountain districts. Nearly all of the gold deposits of the Sierran mines are in quartz veins in granitic rocks, related alaskite and aplite, and rhyolitic dikes. Other metallic minerals, with the exception of iron sulfides and silver minerals, are generally absent. Scheelite is present in a few veins, and galena, sphalerite, chalcopyrite are common in the Cove district.

In 1958, the only continuous gold mining being done was by a group of lessees at the Yellow Aster mine and the only active custom gold mill (stamp and amalgamation) was the Butte Mill at Randsburg.

Through 1957 the recorded production of placer gold in Kern County totaled about 32,000 fine ounces. The actual output of placer gold, however, is probably two or three times larger, partly because, during the early mining of most of the placer deposits, the miners spent the gold as they mined it. In addition, placer gold is not ordinarily sold to the U.S. Mint and its value is not reported.

Most of the placer gold was mined before 1900 by many miners operating individually or in small mining groups. As soon as the easily recoverable gold was mined at one locality, a period ranging from a few months to about 3 years, most of the prospectors moved to other districts or sought lode deposits upstream or upslope from the placer deposits. Since 1900, the only period during which a large number of placer deposits were being worked was during the late 1920s and 1930s.

Although nearly all the gold districts in Kern County have yielded placer gold, the principal placer deposits are in the Rand district, El Paso Mountains, and along Kern River. In general, the placer deposits were discovered and mined before the discovery of nearby lode deposits, and most of the placer deposits have been traced to lode sources. In El Paso Mountains, however, the source of the placer gold has not been found.

The placer gold deposits are most common in Quaternary stream gravels. The gold is commonly most abundant at or near the base of the gravels. Many of the placer deposits were moderately rich in gold and were mined out within a few years after they were found. Some of the gold placer deposits in the Mojave Desert, however, probably could still be worked profitably if they were near an abundant water supply. The high cost of obtaining water to work the deposits, in the and region, by hydraulicking, sluicing, or dredging (in artificial ponds) has hindered the mining of the placer material.

Claraville Placers (Piute Mountain)

Secs. 3,4,5, T29S, R34E, MDM, and vicinity. Piute Mountains area, along Kelso Creek and headwater tributaries from about 1 mile west of Claraville to Landors Meadows about 1-1/2 miles east of Claraville.

Owner: Undetermined (1958); A. V. Fulton, Cantil, operator (1933).

Gravels in creek banks as much as 15 feet high., overlie Mesozoic granodiorite. Fine gold with some small nuggets is present; derived from nearby stringers and quartz veins.

Placer gravels in creek beds and banks worked intermittently for many years by various operators at various places by small scale placer methods. Production undetermined.

Dead River Channel Prospect (Isabella)

Reported in Sec. 17, T26S, R33E, MDM, Isabella area, 2 miles north of Isabella Dam.

Owner undetermined (1958).

A 300-foot-wide gravel bed 6 to 8 feet deep in an east-flowing tributary to Kern River. Gravel rests on granitic rock. Average pay of gravel reported to be 50¢ per cubic yard.

Mostly inundated by Isabella Lake. Mined intermittently since 1890. Production undetermined.

Elbow and Boulder prospect

Sec. 33, T27S, R31E, MDM, on Kern River about 1-1/4 miles northeast of Democrat Springs.

Owner: Harold Bankson, 2007 Cedar St., Bakersfield (1957).

Recent gold-bearing river gravel.

Idle. Most recent reported activity was a sand pumping operation by H. D. Smith, W. M. Cambell, and L. M. Duncan in 1931.

Gold Bar prospect

S1/2 NE1/4 of sec. 33, T27S, R31E, MDM, on Kern River, about 1 mile northeast of Democrat Springs on the northeast tip of a horseshoe bend in the Kern River.

Owner: Mrs. Elizabeth J. Herceg, 114 W. 58th Pl., Los Angeles (1958).

Auriferous Recent gravel. Also contains traces of scheelite.

Intermittent activity since 1934. Produced 1600-1700 cu. yds. of material containing about .05 ounces of gold per cu. yd. Gold-silver ratio 6:1. Idle.

Golden Glow prospect

SE1/4 sec. 31, T26S, R32E, MDM, Greenhorn dist., 2-1/4 miles east of Davis Guard Station.

Owner: George Morrison (1957).

Recent stream gravel. Granitic bedrock.

No recorded production. Idle.

Kelso Creek placers

Sec. 16, T28S, R35E, MDM, Sageland dist., near townsite of Sageland.

Owner: Andrew Miller, Sageland (1933).

Recent stream gravels from 10 to 30 feet thick and 150 feet wide resting on granitic bedrock. Mostly fine gold, but some coarse pieces reported, presumably derived from gold quartz veins in Piute Mountains.

A short-lived attempt was made in 1932 or 1933 to recover the gold by separating fines through revolving screens then passing through sluices. Average gold value of gravels was reported to be 50¢ per cubic yard. Scarcity of water resulted in shutdown.

Keyesville Mines placers

NE1/4 of sec. 35, T26S, R32E, MDM, Keysville dist., 2 miles northwest of Isabella, on the flat, 1/4 mile northeast of old Keysville.

Owner: A. D. Coe, Box 662, Isabella (1957).

Auriferous Recent stream gravels in intermittent streams.

Worked by simple panning methods during periods when streams are running. No recorded production.

Keyesville placer

Reported in sec. 36, T26S, R32E, MDM, east of old Keysville (1919).

Owner: G. Henschkel, Isabella (1919); undetermined, 1958.

Abandoned river channel trending northwest, with about 10 to 15 feet of loose, poorly sorted gravel. Bedrock is Isabella granodiorite.

Probably old name. Production history limited to period 1894-1919, when more than 600 ounces gold were recovered.

Maltby Placer

Sec. 15 T25S, R29E, MDM, 4 miles north-northwest of Woody, 3/4 mile southeast of Isham Hill.

Owner: J. Maltby, Woody (1956).

Auriferous gravel in old channel. Gravel bed 150 to 350 feet wide and four to six feet thick has been prospected for 1,300 feet along its length. It is covered by a two to six-foot-thick mantle of soil and lies on granitic bed rock.

Active intermittently since 1894. Last active 1941. Ore was handled by power shovel and trucked to nearby mill. More than 1,500 cubic yards have been mined which averaged .024 ounce gold per cubic yard.

Mayflower gulch

Reported 7 miles northwest of Havilah (1896); not confirmed (1958).

F. F. Boettler, Woody (1896); undetermined, (1958).

Gold-bearing stream gravel 18 inches to 2 feet thick.

Uncorrelated old name. Worked by ground-sluicing in 1896. Probably long abandoned.

Riveredge placer

Reported in sec. 19, T27S, R32E, MDM, about 8 miles southwest of Bodfish on the Kern River between Greenhorn and Stovepipe creeks (1949), not confirmed (1958).

Owner: P. W. Wilson, P.O. Box 8, Kern Canyon Rt., Bakersfield (1949). Undetermined (1958).

Very fine free gold in Recent gravels of the Kern River. Concentrates said to consist of large amounts of black sand.

Uncorrelated name. Probably abandoned. Originally comprised 8 claims along 3,000 feet of river bank.

Soledad Producers and Leard

Reported in sec. 34, T28S, R28E, MDM, 1 mile northeast of Oil City, along Kern River (1904). Not confirmed (1958).

Reed Crude Oil Co., Los Angeles (1904). Undetermined (1958).

Placer deposit. Uncorrelated old name. Probably long abandoned.

Walker

Reported on Kern River, 4 miles south of Isabella (old site), (1896). Not confirmed, 1957.

Owner: Undetermined (1957).

Gold-bearing gravel 4 feet deep.

Uncorrelated old name. Worked by ground-sluicing in 1896. Water was taken from French Creek in 5-mile-long ditch. Production undetermined.

Willow placer

Reported in NW1/4 sec. 27, T25S, R33E, MDM, Cove dist., 1-3/4 miles south of (new) Kernville, now below the highwater level of Lake Isabella; not confirmed (1958).

Owner: U.S. Army Corps of Engineers (1958).

Auriferous river gravel on schist and granodiorite bedrock. Pay zone said to be 6 feet thick.

A ground-sluicing operation prior to 1914. Long idle. [11]

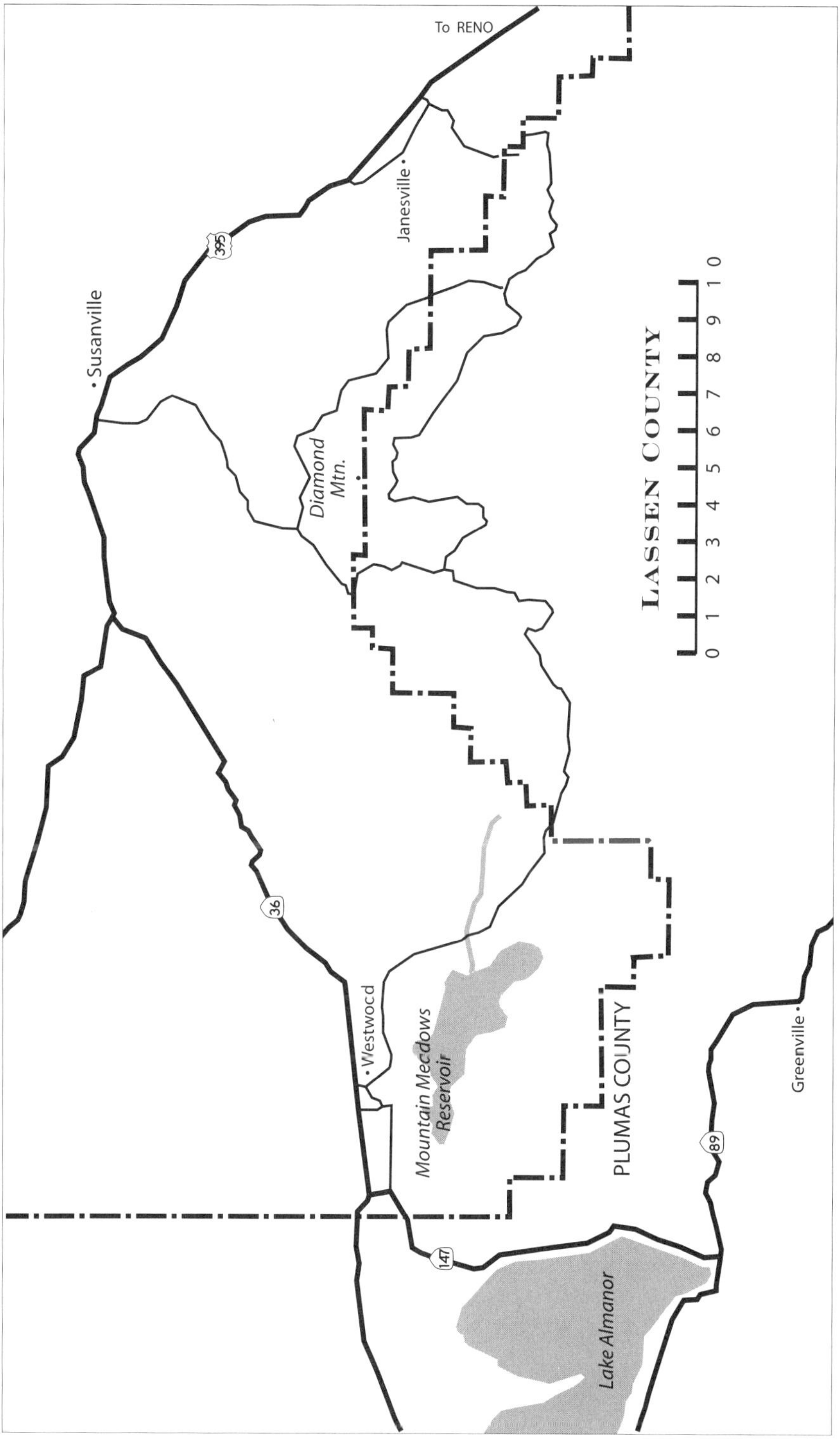
To RENO
Susanville
Janesville
395
Diamond Mtn.
LASSEN COUNTY
0 1 2 3 4 5 6 7 8 9 10
36
Westwood
Mountain Meadows Reservoir
PLUMAS COUNTY
Greenville
89
147
Lake Almanor

LASSEN COUNTY

Diamond Mountain

Location and History. This district is in southern Lassen County about five miles south of Susanville. It is in the Diamond Mountain block, which lies at the extreme north end of the Sierra Nevada. Placer-mining began in the late 1850s, and several hundred thousand dollars in gold were soon recovered. Later some work was done by Chinese miners. Lode-mining began in the 1860s and continued sporadically through the early 1900s with some activity again in the 1930s. The output of the district is estimated at around $1 million.

Geology. The district is underlain by quartz diorite and granite, which is overlain in places by Tertiary gravel and andesite. The quartz veins are as much as 15 feet thick and occur in shear zones chiefly in the quartz diorite. The ore contains free gold and varying amounts of pyrite. Milling ore usually was low to medium in grade, but some high-grade pockets were found. The ore shoots were not large and none of the mines has been developed to any great depth. Some aplitic and basic dikes are associated with veins. Opal also is in the veins.

Mines. Arkansas $200,000+, Gold Belt (McDow) $20,000, Honey Bee $50,000+, Honey Lake, Harris-Mosgrove, Red Jacket (Gayman).

Mountain Meadows

Mountain Meadows is in southwestern Lassen County a few miles southeast of Westwood and eight miles north of Greenville. Years ago placer gold was recovered from Tertiary gravels in the southeast end of the valley. These deposits are on the northwest end of a Tertiary channel known as the Jura channel, which extends southeast to the Taylorsville and Genessee districts. Bedrock consists of greenstone and amphibolite.

MADERA COUNTY

- McKenzie-Minturn Mine. This is a drift mine on the north bank of the San Joaquin River, in Sec. 34, T. 10 S., R. 21 E., 3 miles northeast of Friant. Owners, A. H. McKenzie and W. Minturn, Fresno. Both above and below this point on the south side of the river and above it on the north side, these placer gravels were worked between 1850 and 1860, the camp being known as Millerton, the county seat of Fresno County. The present workings were commenced n 1904. The ground is patented. The gravel is in an ancient river channel with a decomposed granite bedrock. The gravel, which is firm but not cemented, is 3 to 6 feet thick, and the bedrock runs about 28 feet below the surface of the ground. The channel is 5 to 6 feet higher than high-water mark of the present river, and has a different character of gravel, being coarser and made up of different rocks.

  The main adit was driven in 700 feet and stoping began at the back end, the ground being allowed to cave in behind the work. The bedrock shows three channels, the one farthest back from the present river channel carrying the best values. The gold is coarse, but without nuggets, the largest single piece found weighing only $3.20, and is stated to assay better than $19 per ounce. About 200 feet square of the 160 acres have been worked. The gravel is stated to average around $2 per cubic yard. The method of working is to stope out the gravel, fill behind with the coarse material, and tram out the sand and small stuff to the sluices. Sixteen-foot sluices with Hungarian riffles are used. A 6-inch centrifugal pump driven by a 40 h.p. distillate engine furnishes water from the river. Two men are employed (August, 1914). [3]

Chowchilla

In recent years small amounts of placer gold have been mined from the lower Chowchilla River west of Raymond in western Madera County. The gold has been recovered by several small- to medium-sized floating suction dredges, which consist of pontoons that carry suction pumps and gold-recovery equipment. The river gravels range from 10 to as much as 35 feet in depth. This area was first placer-mined during the gold rush, when the Grub Gulch district to the northeast was active.

Coarsegold

Location. This district is in east-central Madera County in the vicinity of the town of Coarsegold about 30 miles northeast of Madera. It is in the central portion of a 20-mile long belt that extends from Grub Gulch on the northwest to Fine Gold on the southeast. It includes part of the area once known as the Potter Ridge district. Coarse and heavy placer gold was recovered from shallow deposits during the gold rush. Lode mining began in 1853, when the Texas Flat mine was discovered. The district has been intermittently mined and prospected ever since.

Geology. The area is underlain by medium to coarse-grained granodiorite with some narrow belts of slate and schist. A number of north-trending quartz veins that contain free gold and varying amounts of sulfides are enclosed in both the metamorphic and granitic rocks. The veins were mined to depths of 1000 feet.

Mines. Baker, Balfron, Daisy Bell, Five Oaks, Golden Road, Melvin, Morning Star, New Citizen, Texas Flat $200,000+.

Fine Gold

Location. This district is in east-central Madera County at the site of the old town of Fine Gold, about 35 miles northeast of Madera. It is in the south end of a 20-mile mineralized belt that extends southeast from the Grub Gulch and Coarsegold districts. It also includes the Quartz Mountain area. The district was first mined in the late 1850s. There has been intermittent prospecting in recent years, chiefly in the vicinity of Quartz Mountain.

Geology. The region is underlain by medium- to coarse-grained granodiorite cut by pegmatite and aplite dikes. Also present are thin beds of gneiss, schist and slate. There are a number of quartz veins containing small but sometimes rich ore-shoots. The ore contains native gold, pyrite, and galena. Nearly all of the deposits are shallow. The veins range from less than one to more than 10 feet in thickness.

Mines. Ackers, Fresno Banner, Johnny, Little Johnny, Quartz Mountain, Standard, Waterloo, Zebra.

Fresno River

Small amounts of gold are recovered intermittently from the lower Fresno River in western Madera County. There are several small floating suction dredges similar to those on the Chowchilla River to the north (see also the section on the Chowchilla district). This stream was first placer-mined during the gold rush, when the Coarsegold and Grub Gulch districts to the northeast were originally worked.

Grub Gulch

Location and History. This district is in east-central Madera County at the site of the old town of Grub Gulch, seven miles north of Coarsegold and 35 miles northeast of Madera. The site is at the northwest end of a 20-mile-long belt that extends from here southeast through the Coarsegold and Fine Gold districts. This district has been the most productive portion of this belt. It includes part of the area that also was known as the Potter Ridge district. The camp was established shortly after the discovery of Coarsegold in 1849. There was much activity from the 1880s through the early 1900s but very little mining has been done since. A few small suction dredges were active in the 1940s and 1950s downstream in the Chowchilla River.

Geology and Ore Deposits. The district is underlain by a northwest-trending belt of mica schist and quartzite with granodiorite to the west. A series of gold-quartz veins are variously oriented. The veins range from one to 10 feet in thickness. The ore contains free gold and varying amounts of sulfides. The milling-grade ore was reported to have yielded up to one ounce of gold per ton. The greatest depth of development is 800 feet.

Mines. Bullion, Butterfly, Canary, Crystal Spring, Enterprise $100,000, Gambetta $500,000, Hoboken, Josephine $360,000, Lucky Bill, Savannah, Starlight, Woodland.

Hildreth

Location and History. This district is in cast-central Madera County at the site of the town of Hildreth about 35 miles east of Madera. It is on the northwest end of an indistinct belt of gold mineralization that extends southeast through the Temperance Flat and Big Dry Creek districts in Fresno County. Apparently the chief period of mining in the area was from about 1860 to 1890, with possibly some prospecting and development again during the 1920s and 1930s. The district was named for the Hildreth brothers, farmers who settled here about 1870.

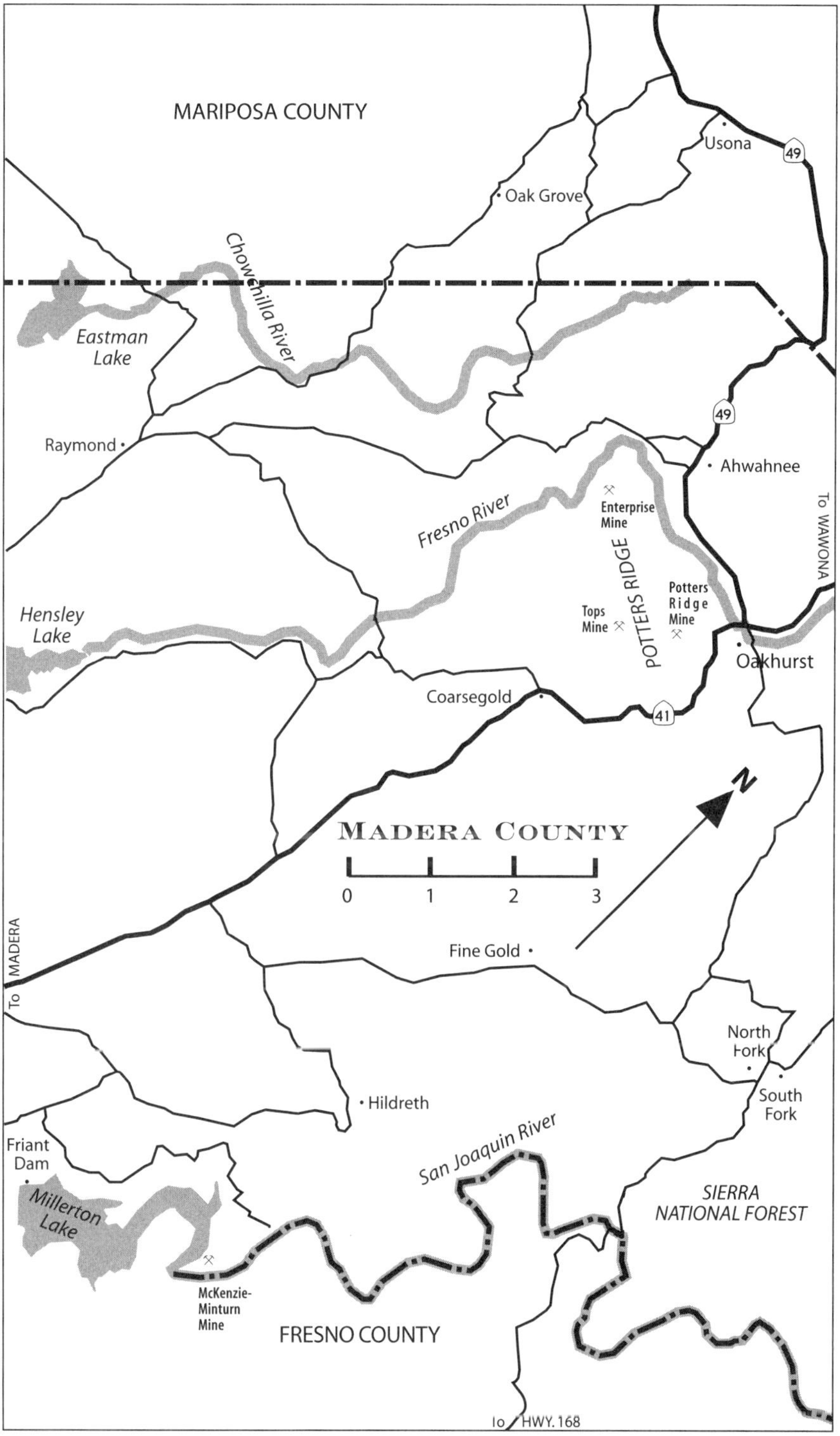
MARIPOSA COUNTY
Usona
49
Oak Grove
Chowchilla River
Eastman Lake
Raymond
49
Ahwahnee
Fresno River
Enterprise Mine
To WAWONA
POTTERS RIDGE
Potters Ridge Mine
Tops Mine
Hensley Lake
Oakhurst
Coarsegold
41
MADERA COUNTY
0 1 2 3
N
To MADERA
Fine Gold
North Fork
South Fork
Hildreth
Friant Dam
San Joaquin River
Millerton Lake
SIERRA NATIONAL FOREST
McKenzie-Minturn Mine
FRESNO COUNTY
To HWY. 168

Geology. The principal rock in the district is medium- to coarse-grained granodiorite with several narrow northwest-trending beds of slate and schist. There are a number of north-striking quartz veins containing free gold and often abundant sulfides. The veins are as much as 20 feet thick, and several have been developed to inclined depths of about 600 feet.

Mines. Abbey $100,000+?, Golconda, Hanover, Hildreth, Morrow (Moro, Bazinet), Mud Springs $250,000, Volcano No. 1 $100,000.

MARIPOSA COUNTY

• Placer mining in the county has dwindled to almost nothing and what little work is being done is very irregular and depends very extensively on the rainfall. One does not see the long ditch lines in this county such as are to be found in other counties, bringing water to the placer mines. [3]

• MARIPOSA PLACER MINES—GENERAL REVIEW

Although the placers of Mariposa County yielded millions of dollars to the early miners, they consisted chiefly of creek deposits that were both shallow and narrow. In consequence, they were quickly exhausted. In some localities the placers were fairly rich, but they never approached the bonanza gravel deposits of Tuolumne and Calaveras Counties either in richness or yield. No remnants of rich Tertiary placers such as exist in Calaveras and Tuolumne Counties have been preserved, as the protective blanket of lavas and other ejecta from the great volcanic eruptions of the late Tertiary that filled many stream valleys to the north did not extend as far south as Mariposa County. Hence, the Tertiary gravels of the county were swept away, and all of the existing placers may be classified as Quaternary.

The principal placer localities of the past were the headwaters of Mariposa Creek in the vicinity of Mariposa and Mormon Bar and its tributaries, Agua Fria, Buckeye, and Stockton Creeks; Bear (west of Cathay), Corbett, Hornitos, and Burns Creeks; and the bars and tributaries of the Merced River west of Clearing House, including the North and South Forks and Sweetwater, Feliciana, Bear (tributary to the Merced at Briceburg), Saxon, Sherlock, and Maxwell Creeks.

There are no authentic records of early placer production, but it probably did not exceed $10,000,000. Virtually all of this was produced prior to 1870, by which time all the richer deposits had been worked out. Although accurate records of the production of placer gold in the county before 1903 are not available, it is improbable that this exceeded 1,000 ounces in any year between 1870 and that date. The average annual output between 1903 and 1934 was 220 ounces, with a maximum of 603 ounces in 1933. However, under the stimulus of a higher price for gold, production increased to 1,100 ounces in 1934; and with the installation of several draglines and floating washing plants that are working small acreages of gravel too low in grade to have been profitable heretofore, production increased to 7,137 ounces in 1937 and probably exceeded 10,000 ounces in 1938. Any further increase in the placer production can be only temporary and it is inevitable that within a few years the output will decline as it again becomes dependent on small sporadic operations that characterized it for so many years. It is of interest to note that several hundred persons, including many itinerants, were engaged in "sniping" for placer gold in this county during 1938. The "sniper" is generally one who owns no ground but moves about from place to place to work small bits of virgin ground by hand for a short time, or to dig in crevices of exposed bedrock for a little gold brought down by rains of recent years. At several places, notably on the headwaters of Agua Fria and Bear creeks, small camps of 20 to 30 shacks, tents, and trailers were established by the snipers. Their earnings, which generally average less than $1 a day, were increased somewhat in 1938 by the unusually heavy rains of the preceding winter, which reconcentrated many shallow deposits and in streams with steep gradients exposed considerable areas of bedrock for crevicing.

In the summer of 1938 three draglines and floating washing plants were in operation in the county, all on Mariposa Creek. One, belonging to the Trebor Corporation, was mining at Mormon bar, and the other two, owned by Placer Properties, Inc., and the K. R. Nutting Co., respectively, were operating about 20 miles farther downstream near the western boundary of the county. Both companies last named expected to work out their ground in 1938, and the Trebor Corporation had only a small yardage left at Mormon bar. Although there are a few other small areas in the county that may be worked profitably by draglines, the steep gradients of the creeks and the generally hard and unevenly eroded bedrock do not favor this type of mining. It is evident, therefore, that the companies now operating draglines in the county soon will be forced to seek elsewhere for workable deposits and, on the whole it seems improbable that even the present small production from placer mining in Mariposa will be maintained for more than a few years.

Brief descriptions of the principal placer operations in the county in 1938 follow.

DRAGLINE OPERATIONS

PLACER PROPERTIES, INC.—In March 1936, Placer Properties, Inc., of which R. G. Kumle, Oregon House, Calif., is president, placed a dragline washing plant in operation on a 125-acre tract on Mariposa Creek, about 8 miles by road east of Le Grand in sec. 31, T. 7 S., R. 17 E. L. Brier of Le Grand is superintendent

The pay gravel extends along the creek and has an average width of 600 feet and depth of 14 feet. The gravel is loose and medium size and contains no clay and only occasional bed boulders. The gold content averages about 15 cents per cubic yard The gold is chiefly fine, although some individual pieces worth as much as 25 cents are recovered occasionally. The bedrock at the lower end of the property is diorite, and upstream this is succeeded first by granite and then by slate.

About 100,000 cubic yards of gravel is excavated and washed per month. When visited in the summer of 1938 the tract was nearly worked out, and it was expected that the operation would be concluded in October of that year. The recovery is reported as about 13 cents per cubic yard.

In February 1938 the K. R. Nutting Co. of Le Grand began mining with a dragline and floating washing plant on a 70-acre tract on Mariposa Creek about 4 miles east of Le Grand in sec. 1, T. 7 S., R. 16 E. and just within the western boundary of Mariposa County. T. J. Booth of Le Grand is superintendent.

The pay gravel follows the general course of the creek and averages 700 feet in width and about 13 feet in depth but has a maximum width of 1,000 feet. Its gold content is said to average, about 15 cents per cubic yard.

When the property was visited in August 1938, the, stripping of about 180,000 cubic yards of overlying soil with an average depth of 3-1/2 feet was in progress.

About 2,800 cubic yards of gravel per day was being excavated and washed, with a reported average recovery of about 13 cents per cubic yard. Including the superintendent, 13 men were employed each day, the plant being operated three shifts with three men per shift and the additional services of a welder, a roustabout, and bookkeeper on the day shift. It was expected that this particular tract would be worked out by January 1, 1939.

TREBOR CORPORATION—During the summer of 1933 the Trebor Corporation, of which Robert D. Mueller of Mariposa is president and general manager, acquired placer leases

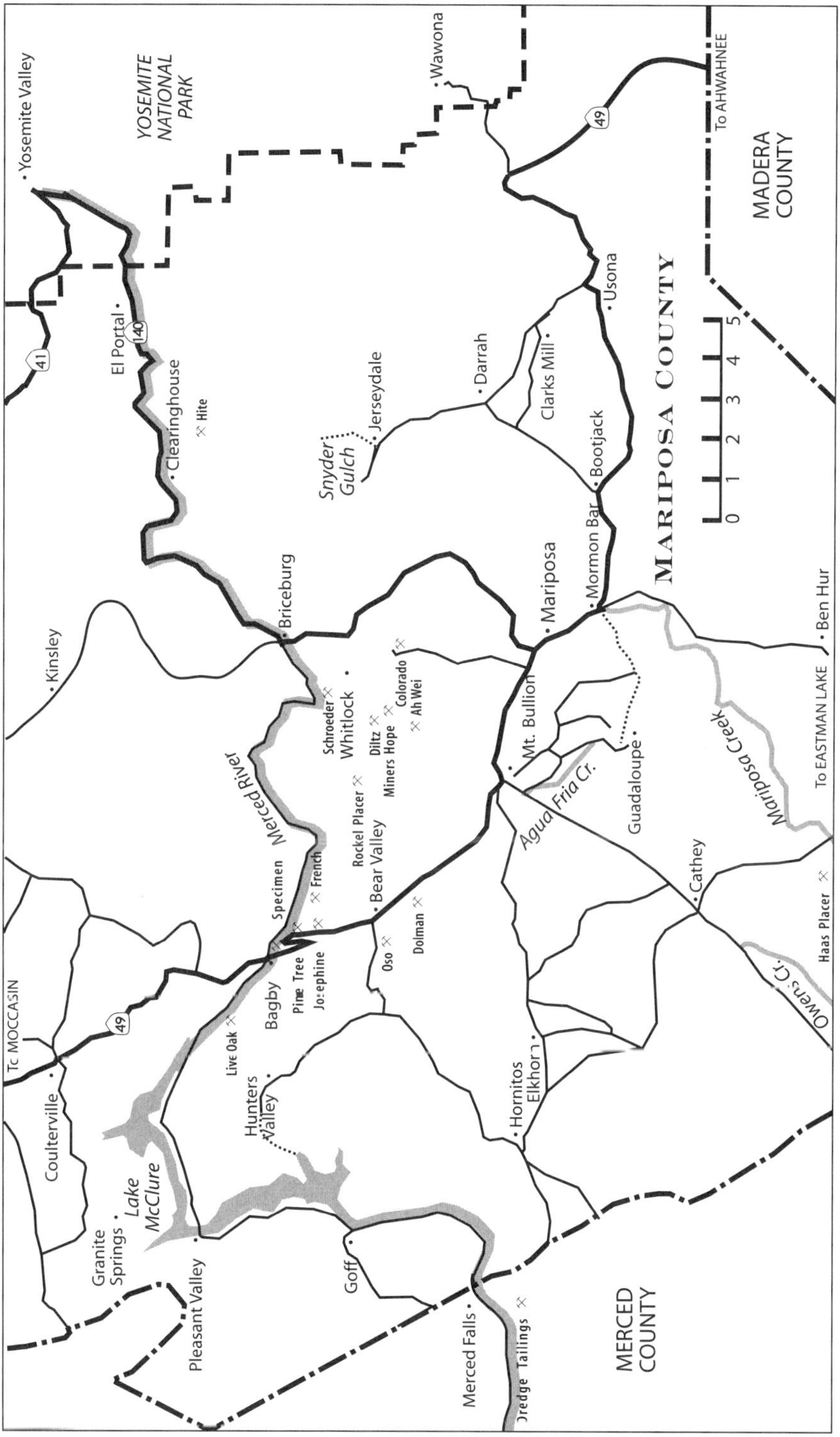
MARIPOSA COUNTY
YOSEMITE NATIONAL PARK
MADERA COUNTY
MERCED COUNTY
Yosemite Valley
Wawona
To AHWAHNEE
El Portal
Clearinghouse
Hite
Jerseydale
Snyder Gulch
Darrah
Clarks Mill
Usona
Bootjack
Mormon Bar
Mariposa
Briceburg
Kinsley
Merced River
Schroeder
Whitlock
Colorado
Ah Wei
Diltz
Miners Hope
Mt. Bullion
Guadaloupe
Agua Fria Cr.
Mariposa Creek
Ben Hur
To EASTMAN LAKE
Rockel Placer
Bear Valley
Specimen
French
Dolman
Oso
Josephine
Pine Tree
Bagby
Live Oak
Cathey
Haas Placer
Owens Cr.
To MOCCASIN
Coulterville
Hunters Valley
Hornitos
Elkhorn
Lake McClure
Granite Springs
Pleasant Valley
Goff
Merced Falls
Dredge Tailings
0 1 2 3 4 5

on Mariposa and Agua Fria Creeks from the Mariposa grant and installed a dragline and floating washing plant on the former creek at Mormon bar.

Twelve men are employed in the operation of the plant, and three more constitute a crew that is engaged in sampling gravel deposits held under lease. Mining was begun in September 1938, and 1,600 cubic yards per day was handled until the available yardage was worked out early in 1939, when the plant was moved to another placer on Agua Fria Creek.

The average depth of the gravel at Mormon bar was only 6 feet, and mining was rendered difficult by many large boulders and bad bedrock conditions. Pinnacles of bedrock that often projected above the water level gave trouble in maneuvering the boat. The management stated that maintenance and mining costs were excessive, owing to the rough operating conditions, and that actual running time averaged only about 16 hours per day. A recovery of 22 cents per cubic yard is reported.

HANAS PLACER—The T. D. Hanas placer lies in the steep V-shaped valley of Mariposa Creek in sec. 9, T. 7 S., R. 17 E., about 1-1/2miles northwest of White Rock and 8 miles downstream from Mariposa. It contains 40 acres and extends about 6,000 feet along the creek. The gravel is 9 to 25 feet deep and is said to average more than 50 cents per cubic yard. The bedrock is an unevenly eroded slate.

In 1937 a partnership was formed to mine the placer with a dragline and floating washing plant, but difficulties arose that resulted in removal of the dragline from the property. When visited in August 1938 the plant had no excavating machinery, and the washing unit was of necessity idle. Most of the gold in this deposit is fine, but pieces worth 5 to 10 cents each are common and nuggets valued up to $2 are found occasionally. Hanas was negotiating for a dragline and hoped to operate the plant during the winter of 1938.

MISCELLANEOUS OPERATIONS

KOCKEL PLACER—A creek placer that has been covered by a slide is being developed by Traugott Kockel on the south slope of Sherlock Gulch in sec. 24, T. 4 S., R. 17 E., in the Sherlock mining district. A few pounds of exceptionally heavy gold has been taken out, with nuggets weighing up to 1-1/2 ounces, but development is as yet in the preliminary stage, all work having been done by hand. [7]

• AH WEI DRIFT MINE—Consists of one unpatented claim (placer) owned by John Hand and W. D. Weston of Mariposa and located in Sec. 32, T. 4 S., R. 18 E., M.D.M., 25 miles from Bagby. The gravel is bench gravel with a rough slate bedrock with free wash and large boulders.

The development consists of 2000 feet of bedrock tunnel. The equipment consists of 2 miles of ditch, a reservoir 30 feet by 10 feet by 4 feet deep, sluice boxes, and pole and Hungarian riffles. There are cabin, blacksmith shops, drills and tools and one ore car. Works six months during the winter season.

SCHROEDER PLACER MINE—Consists of one patented claim owned by J. A. Schroeder, P. W. Judkins, and C. H. Weston and located in Sec. 16, T. 4 S., R. 18 E., M.D.M., 20 miles from Bagby. These are bench diggings, and the gravel is medium wash with some clay and boulders on a porphyry bedrock. The development consists of about 5 acres that have been washed off by ground sluicing with a canvas hoe. A tunnel 50 feet long is being extended. The equipment consists of 3 ditch lines, having a total length of 5-1/2 miles, 1500 feet of 8-inch pipe, sluice boxes and riffles. It is being worked each winter. [3]

Bagby

Location. This district is in western Mariposa County in the vicinity of the towns of Bagby and Bear Valley. It is in the Mother Lode gold belt.

History. The streams were placer-mined early in the gold rush, and the Pine Tree and Josephine Veins were discovered in 1849. Part of the area was on the Las Mariposa Spanish land grant of General John C. Frémont. Bagby was first known as Benton Mills, named by Frémont for Senator Thomas Hart Benton, his father-in-law. It was renamed in the 1890s for A. Bagby, a hotel owner. The town was a stop on the Yosemite Valley Railroad, which once extended up the Merced River canyon to Yosemite National Park.

Gold mining activity continued until around 1875. There was mining in the district again in the early 1900s. The Pine Tree-Josephine mine was worked on a major scale from 1933 to 1944, and the Red Bank mine has been active in recent years. Part of the area, including the old town of Bagby, was inundated by the Exchequer Reservoir in 1967.

Geology. In this district the Mother Lode gold belt is about 1-1/2 miles wide. It is underlain by northwest-striking beds of slate, phyllite, and metasandstone of the Mariposa Formation (Upper Jurassic), with greenstone and green schist both to the west and southeast. A belt of serpentine extends northwest through the central portion of the district and is structurally important in relation to some of the gold-bearing veins.

Ore deposits. There are several northwest-trending vein systems that consist of quartz veins and stringers with brecciated slate, schist, and associated bodies of pyritic ankerite and mariposite-quartz rock. These vein systems are often scores of feet in thickness. The ore contains free gold, pyrite, and arsenopyrite with small amounts of chalcopyrite, galena, millerite, sphalcrite, and niccolite. Milling ore averaged 1/7 to 1/2 ounce of gold per ton. In places high-grade ore is abundant. The ore shoots had stoping lengths of up to 600 feet, and the veins were mined to an inclined depth of 1500 feet.

Mines. Dolman, French $116,000+, Jumper, Juniper, Live Oak, Mexican 1 $50,000, Oso $50,000+, Pine Tree-Josephine $4 million+, Queen Specimen, Red Bank $100,000+, Specimen.

Cathey

The Cathey or Cathay district is in southwestern Mariposa County about 10 miles southwest of Mariposa and near the town of Catheys Valley. The area was first placer-mined during the gold rush, and lode mining began in the 1850s. The principal mines were the Francis, Moore Hill, and Rich mines, which were last worked in the 1930s. A number of north-trending quartz veins up to 9 feet thick, in granodiorite and schist, contain native gold and sulfides. The sulfides, which include pyrite, arsenopyrite, and galena, sometimes are extremely abundant.

Cat Town

This district is in northwestern Mariposa County at the site of the old mining camp of Cat Town. Although the Kinsley-Greeley Hill district adjoins it on the north and the Coulterville portion of the Mother Lode is on the west, this district is on a separate northwest-trending vein system. It may be on the same belt of mineralization as the Whitlock district to the southeast. The height of activity here was in the 1880s and 1890s. The Bug mine and several others were active in the 1930s and there has been some prospecting since. Also the area was recently prospected for molybdenum.

The deposits consist of gold-bearing quartz stringers in schist, slate, metachert, and greenstone. The values usually occur in small but rich pockets and are closely associated with albitite and diorite dikes. The deposits have not been mined to depths of greater than 100 feet. The principal gold sources have been the Black Bart, Gold Bug, and White Porphyry mines.

Clearinghouse

Location and history. Clearinghouse is in central Mariposa County a few miles west of El Portal and Yosemite National Park. It was named for the Clearinghouse mine, the largest source of gold in the district. The Merced River, which flows through the area, was extensively placer-mined during the gold rush. The Clearinghouse mine was discovered in 1860 and worked on a fairly large scale until about 1880. There was mining activity again during the early 1900s and 1930s, and there has been intermittent prospecting and development work since. Substantial quantities of limestone and barite and some tungsten are found here. At one time the Yosemite Valley Railroad extended through the area to El Portal, the line's eastern terminus. From El Portal, passengers were taken by stage to Yosemite Valley.

Geology and ore deposits. The principal rocks underlying the district are graphitic schist, slate, quartzite, and hornfels of the Calaveras Formation (Carboniferous to Permian). There are some granitic dikes, and the main mass of the Sierra Nevada granitic batholith is a few miles to the east.

The gold deposits occur in north-striking quartz veins that usually range from one to five feet in thickness. The ore contains free gold and often abundant sulfides. Appreciable amounts of high-grade ore have been found here, and milling-grade ore commonly contained one ounce or more of gold per ton. Several of the veins have been developed to inclined depths of about 1200 feet.

Mines. Clearinghouse $3.35 million, Gold Star, Old Timer, Rutheford and Cranberry, South Cranberry Uncle Jim, West Rutherford.

Coulterville

Location. This district is in Southwestern Tuolumne County and northwestern Mariposa County. It is that portion of the Mother Lode gold belt that extends from the vicinity of the McAlpine mine southeast through Peñon Blanco and the town of Coulterville to Virginia Point, a distance of about 10 miles.

History. The surface portions of the veins and the streams were worked during the gold rush. The town was named for George Coulter, who opened a store there in 1849. Quartz-mining began about 1852 with the discovery of the Malvina and Mary Harrison veins. Considerable lode mining was done from the 1860s through the 1890s, when many of the mines belonged to the Cook estate. There was some activity from the early 1900s until 1942. In recent years some work has been done at the Virginia, McAlpine and a few other mines.

Geology. In this district, the Mother Lode system is associated with an extensive northwest-trending lenticular body of serpentine. In a number of places the serpentine has been hydrothermally altered to mariposa-quartz-ankerite rock. Slate and greenstone lie to the west, and amphibolite schist and greenstone are to the east. Also present are smaller amounts of chlorite schist, phyllite, and metadiorite and aplitic dikes.

Ore Deposits. The ore deposits consist of pyrite-bearing bodies of mariposite-quartz-ankerite rock, with numerous parallel quartz stringers and adjacent massive quartz veins.

The gold occurs in the free state or with the pyrite and usually is found in the quartz stringers, although in places there are values in the massive veins. A number of high-grade pockets have been found in this district, some of which came from the gossan that overlies some of the deposits. Tellurides also have been found here. Milling ore usually ranges from 1/5 to 1/2 ounce of gold per ton. The ore shoots had stoping lengths of up to 400 feet and were mined to inclined depths of more than 1200 feet. The massive veins form prominent ridges; Peñon Blanco Ridge is visible for many miles.

Mines. Mariposa County: Adelaide, Big Lode, Champion 1 $150,000 to $200,000, Flyaway group, Louisa $100,000, Malvina $1 million, Mary Harrison $1.5 million, Midas, Oro Rico, Potosi, Tyro $110,000+, Virginia $824,000. Tuolumne County: McAlpine $100,000+.

Granite Springs

This district is in the northwest corner of Mariposa County and the southwest corner of Tuolumne County in the vicinity of Lake McClure and Don Pedro Reservoir. The mines apparently were last worked in the 1930s. The region is underlain chiefly by greenstone with some interbeds of graphitic slate. There are a considerable number of northwest-striking gold-quartz veins often associated with diorite dikes. Most of the deposits are relatively shallow.

Mines. Mariposa County: Anita, Burr $350,000, Florinita, Jackson, White Rock. Tuolumne County: Buzzard Roost, Diamond, Hedley, Oak Mesa, Solambo.

Gravel Range

This district is in south-central Tuolumne County and north-central Mariposa County about 15 miles east of Groveland. The principal sources of gold have been several bodies of quartzitic Tertiary channel gravels that are part of the Tertiary Tuolumne River. The deposits are in the Gravel Range, at Dorseys and north of Smith Station. They were mined chiefly by hydraulicking. Bedrock consists of granodiorite in the east and slate and schist in the west. In places the gravels are capped by andesite.

Hite Cove

Location and History. Hite Cove is in central Mariposa County on the South Fork of the Merced River. Placer mining began in the area shortly after the beginning of the gold rush, and the Hite mine was discovered in 1862 by John R. Hite. He operated the property for 17 years and became quite rich. The mine was active again during the early 1900s. There has been some prospecting in the area in recent years.

Geology and Ore Deposits. The district is underlain by graphitic schist and slate, quartzite, and hornfels. These rocks are cut by a variety of aplitic and granitic dikes, some of which are associated with goldquartz veins. There are a number of northwest-striking quartz veins up to 12 feet thick. The ore contains native gold and often abundant sulfides. The greatest depth of development is about 800 feet.

Mines. Brown Bear, Bunker Hill, Confidence, Emma, Eureka, Georgia Point, Hite $3 million, Hite Central, Kaderitas, Mexican, Williams.

Hunter Valley

Location and History. This district is in the northwest corner of Mariposa County, in the general area of Hunter Valley, the Don Pedro Reservoir and Lake McClure. It was named for William W. Hunter, a well-known engineer. There was extensive placer gold mining here during the 1850s and some copper mining in the 1860s. The lode gold mines were active until the early 1900s. Some mining was done again during the 1930s, and the Pyramid mine has been prospected recently.

Geology. The district is underlain by northwest-trending belts of slate of the Mariposa Formation (Upper Jurassic) and greenstones, chert and slate of the Amador Group (Middle to Upper Jurassic). Several small diorite and granodiorite intrusions are mapped.

Ore Deposits. A number of northwest-striking systems of gold-quartz veins are in the slate, chert and greenstone. Numerous stringers and cross veins are present. A number of high-grade pockets have been found. The ore contains free gold and often abundant sulfides, and mining ore commonly averaged one ounce per ton in gold. None of the veins have been mined to depths of more than a few hundred feet.

Mines. Blue Cloud, Cotton Creek, Iron Duke, Morning Star, Oak and Reese $500,000-$600,000, Orange Blossom, Pyramid $200,000, Schoolhouse, Yellowstone.

Jerseydale

Location and History. The Jerseydale district is in west-central Mariposa County. It is about 10 miles northeast of the town of Mariposa and just east of Whispering Pines. It includes the Feliciana Mountain area. The streams were first mined during the gold rush, and the lode deposits were discovered soon afterward. Much activity continued from the 1870s until the early 1900s. Some mining was done again during the 1930s, with intermittent prospecting since.

Geology. The principal rocks in the area are slate, graphite schist and phyllite of the Calaveras Formation (Carboniferous to Permian) and granitic rocks. Also present are greenstone and dioritic dike rocks. Numerous narrow quartz veins contain small but rich highgrade shoots and "pockets". Sulfides are abundant.

Mines. Blue Bell, Buffalo, Comet, Early Feliciana $159,000, King Solomon, Louisa, Monte Christo, Roma $100,000.

Kinsley

Location and History. This is an extensive district in the Sierra Nevada east gold belt in north-central Mariposa County. It is five miles east of Coulterville and about 25 miles north of Mariposa. The district includes several places that at various times have been classified as separate districts but are grouped together in this publication because they adjoin each other and are geologically similar. These are the Greeley Hill, Bull Creek, Gentry Gulch, Smith Ridge, and Dogtown areas. The Cat Town district lies just to the south and the Coulterville district to the west. The area was placer-mined during the gold rush. Much lode mining was conducted from the 1860s through 1900 and again in the 1930s. Several mines including the Hastoe and Horseshoe have been worked intermittently in recent years.

Geology. The district is underlain by slate, mica schist, quartzite, hornfels, phyllites and limestone lenses of the Calaveras Formation (Carboniferous to Permian). Present are several small granitic stocks and a number of diorite, quartz-diorite and aplite dikes. In places these dikes are associated with the gold-quartz veins and are important in the localization of the ore bodies.

Ore Deposits. Numerous north- and west-striking quartz veins range from one to five feet in thickness. The ore shoots generally are small, but commonly rich. The ore contains free gold and often abundant sulfides including galena and tetrahedrite, which are associated with rich ore. Molybdenite is present in a few places.

Mines. Argo $18,000, Bandarita $1.52 million, Bob McKee, Bondurant $390,000, Bunce, Carrie Todd, Contention, Cranberry, Garibaldi, Gold King, Hasloe $3 million, Horseshoe, Last Chance, Lovely Rogers, Louisiana, Marble Springs $200,000, Moonlight, Quail $400,000, Red Cloud, Red Mountain, Texas Hill $74,000+.

Mormon Bar

Mormon Bar is in south-central Mariposa County about three miles south of the town of Mariposa. The area was placer-mined during the 1850s and 1860s, and, by 1870, the easily worked placers were largely exhausted. The area was mined again in the 1930s by dragline dredges, and there has been minor prospecting since. The total gold production for the district is estimated at 75,000 ounces. The deposits are in and adjacent to Mariposa Creek. The average depth of the mined gravels was about six feet.

Mount Bullion

Location. The Mount Bullion district is in west-central Mariposa County about seven miles northwest of Mariposa. The district is in the southern end of the Mother Lode gold belt and extends northwest towards Bagby and Bear Valley. It includes the Agua Fria and Mount Ophir areas.

History. This region was first placer-mined in 1848, many of the miners having been of Spanish descent. Agua Fria Creek and other streams were highly productive (Agua fria means cold water in Spanish). Lode gold-mining began shortly afterward. Much of this district is in the Las Mariposa land grant, which originally belonged to General John C. Frémont. The mines in this grant were not located and surveyed in the same fashion as those on public lands, and to this day the land plats within this grant are difficult to coordinate with established survey lines. The grant later underwent lengthy litigation, and Frémont eventually went bankrupt. He named nearby Mount Bullion for his father-in-law, Senator Thomas Hart Benton, who was sometimes known as "Old Bullion".

At Mount Ophir, which is now a ghost town, are the ruins of an early-day mint. From 1849 until 1854 private coinage subject to federal inspection was authorized in California. It is believed that some of the now extremely rare and valuable octagonal fifty-dollar gold slugs were minted here from locally mined gold.

Gold mining in the district continued fairly steadily from the 1850s through the 1870s. There was considerable activity from around 1900 to 1920 when the Princeton and other mines were worked. Some mining was done in the 1930s and early 1940s, and there have been a few intermittent small-scale operations since.

Geology. The gold mineralization is confined chiefly to a northwest-trending belt of slate, phyllite, and metasandstone of the Mariposa Formation (Upper Jurassic). Within this formation are two belts of pyrite-bearing metarhyolite that may have possible future economic significance. Greenstone of the Peñon Blanco Formation (Upper Jurassic) crops out to the east and west. Also present are thin bands of serpentine and numerous aplite dikes.

Ore Deposits. Several north-northwest-striking systems of quartz veins occur principally in slate. The veins usually range from four to 10 feet in thickness, although there are some massive ones that are considerably thicker. The ore contains free gold and pyrite, which is abundant in places. Milling ore yielded from 1/4 to 1/2 ounce of gold per ton, and considerable highgrade ore was recovered close to the surface. Some of the ore shoots were extensive; several in the Princeton mine had stoping lengths of more than 500 feet. The greatest depth of development is 1600 feet on the incline.

A number of extensive vein systems have not been thoroughly explored. Also, there are several extensive deposits of pyritic metarhyolite in the Mariposa Formation that in places contain gold. These bodies are several miles long and 60 or more feet thick.

Mines. Greens Gulch $119,000+, King Midas, Louis, Mt. Ophir $250,000 to $300,000, Mountain View 1, Nellie Kahoe, Ortega, Princeton $5 million, Sorrel.

Whitlock

Location. The Whitlock district is in west-central Mariposa County five miles north of the town of Mariposa. The district is east of the Mother Lode gold belt and includes the Colorado, Sherlock Creek, and Whiskey Flat areas. The area was placer-mined soon after the beginning of the gold rush, and lode mining began shortly afterward. A number of mines were active here during the 1930s, and a few, such as the Diltz and Schroeder mines, have been intermittently prospected in recent years.

Geology. Greenstone and green schist underlie much of the district, with some slate, phyllite, and mica schist in the north portion. Granitic intrusives and serpentine are to the south. There is an appreciable number of diorite, quartz-diorite, and aplite dikes that commonly are associated with the gold-quartz veins. A northwest-trending fault extends along the west side of the district.

Ore Deposits. Numerous north- and northwest-striking quartz veins contain small but rich ore shoots. The veins usually are one to five feet thick, and a number dip at low angles. The veins have a tendency to roll or bend, and it is in these bends or rolls that the high-grade pockets often occur. Much specimen ore has been produced in the district; in 1932 the Diltz mine yielded 52- and 40-pound masses of gold and quartz. The greatest depth of development is about 900 feet.

Mines. Buffalo, Champion, Colorado $50,000, Diltz $750,000 to $1 million, Geary, Golden Key $154,000+, King Solomon, Landrum, Nutmeg $180,000+, Our Chance, Permit, Schroeder $200,000 to $300,000, Spread Eagle $425,000, Whitlock $500,000.

MERCED COUNTY

• The only gold mining that is being done at present is that at Snelling on the Merced River where the Yosemite Gold Dredging and Mining Company operates a dredge on their 400 acres along the river. The gravel is a clean wash and quite free from clay or large boulders. The gravel averages 20 feet to bedrock and a total of 24,200 cubic yards averaged 16-1/2 cents per yard, and working costs were 6-1/2 cents per yard. [3]

Snelling

Location and History. The Snelling district is in eastern Merced County along the Merced River between the towns of Snelling and Merced Falls. It is principally a dredging field. Some placer mining and hydraulic mining of the terrace deposits along the river were practiced during the gold rush. The town, named in 1851 for Charles Snelling, who operated a hotel and ranch here, was the governmental seat of Merced County from 1857 until 1872. Gold dredging began in 1907 and continued until 1919. There was dredging again from 1929 until 1942 and 1946 to 1952. The value of the total output of the district is unknown, but the dredges are estimated to have produced about $17 million.

Geology. The values were recovered from stream gravels and flood plain and terrace deposits in and adjacent to the Merced River. The gravels are loose with very little clay and range from 20 to 35 feet in depth. The dredged area is about nine miles long and 1/2 to 1-1/2 miles wide. Bedrock is slate in the east and volcanic ash in the west. The gold is fairly fine and about 890 in fineness. A small amount of platinum is present. Dredge recoveries ranged from 10 to 30 cents in gold per yard, with the average close to 10 cents.

Dredging Concerns. Merced Dredging Co., 1932 and 1945-49, 1 dredge; San Joaquin Mining Co., 1936-42, 1 dredge; Snelling Gold Dredging Co., 1932-42 and 1946-52, 2 dredges; Yosemite Mining & Dredging Co., 1907-19, 1 dredge; Yuba Cons. Goldfields, 1930-41, 2 dredges.

• Merced Dredging Company, San Francisco, operated a bucket dredge half a mile south of Snelling in 1940, 1941, and part of 1942.

San Joaquin Mining Company, San Francisco, operated a bucket dredge 3 miles west of Snelling in 1940, 1941, and part of 1942.

Snelling Gold Dredging Company of Snelling operated two bucket dredges on the Merced River between Merced Falls and Snelling throughout 1940 and 1941, and during part of 1942. The property is in secs. 10, 11, 12, T. 5 S., R. 14 E., and covers a deposit of uncemented gravel and sand 10 to 20 feet deep with bedrock of volcanic ash.

Yuba Consolidated Gold Fields, San Francisco, operated a bucket dredge 4 miles east of Snelling in 1940 and part of 1941, in T. 5 S., R. 15 E. Gravel ranged in depth from 18 to 36 feet, and bedrock is slate near Merced Falls but is volcanic ash on the lower part of the tract. A little platinum was recovered with the gold. [1]

• There is only one gold dredging company in Merced County, the Yosemite Mining and Dredging Company, which began operations late in 1907, and has an operating plant of one dredge on the Merced River, near the town of Snelling. The holdings of this company comprise an area of about 400 acres, located along the Merced River, and consisting of river bottom and low land, periodically submerged and therefore not cultivated. The gravel is a clean wash, carrying practically no clay, and no boulders larger than can be handled by a modern dredge. It averages in depth to bedrock about 20 feet, and in places is overlain by a sandy loam several feet in depth. The gravel is quite loose and the water-level close to the surface. Attempts to reach bedrock by sinking shafts were unsuccessful, and the property was prospected by means of a Keystone drill. According to the general manager's report the actual average of 27 test holes sunk over an area of 178 acres was 19-1/2 cents per cubic yard, and the average from 50 holes and pits combined was 16.4 cents per cubic yard. On a tract of 113 acres 20 shafts were sunk to water-level and 26.7 cubic yards of gravel washed, which yielded 14.4 cents per cubic yard. On this same tract 16 drill holes gave an average 23.4 cents per cubic yard contained in the gravel below water-level. Pits along the river bank are said to have yielded as high as 40 cents per cubic yard. The general manager estimates the cubical contents of this land to be 24,200 cubic yards per acre, the average value of the gravel 16-1/2 cents per cubic yard, and the cost of dredging at 6-1/2 cents per cubic yard. The gold is said to be relatively coarse, clean, and easily saved. [3]

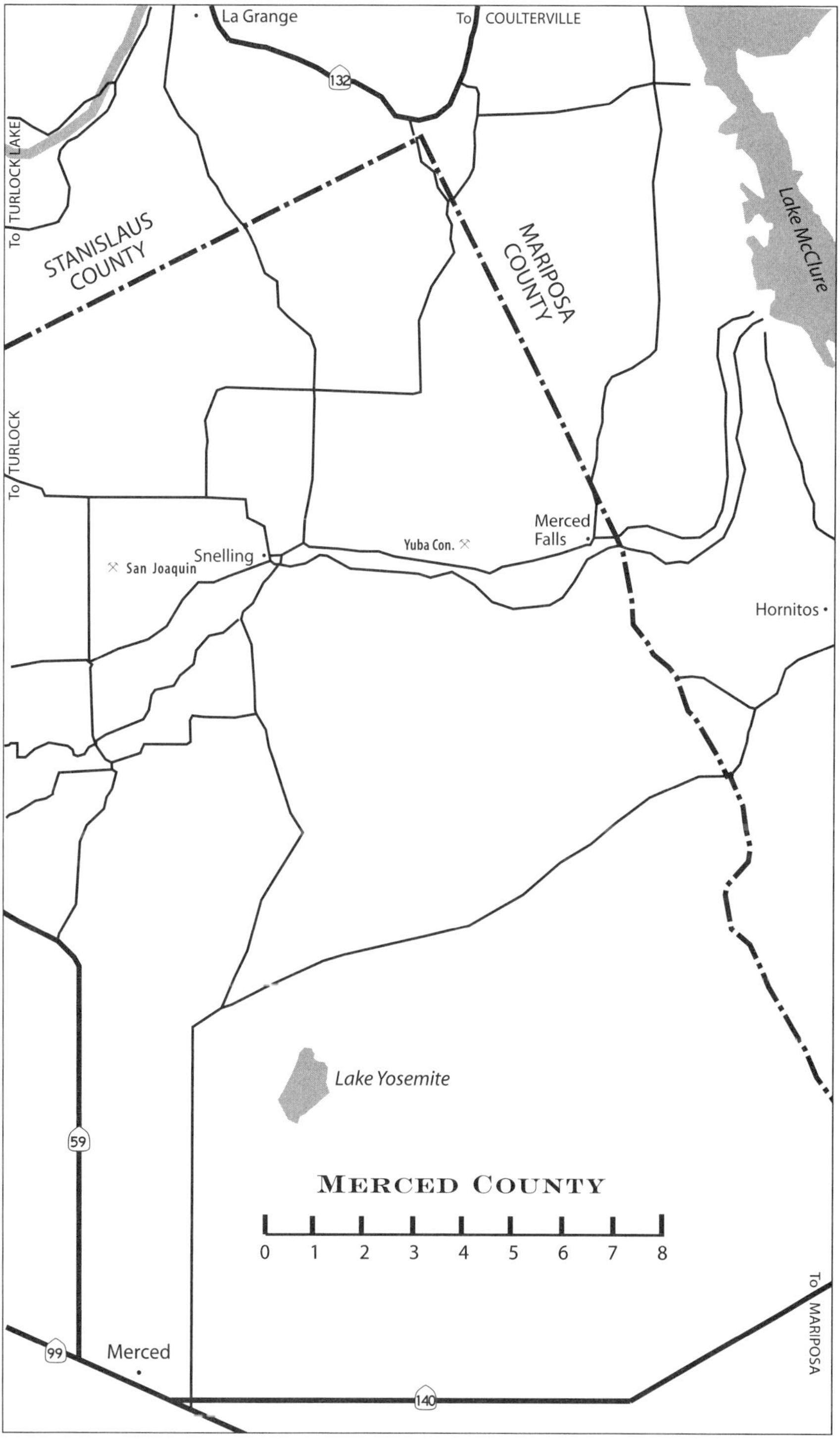
La Grange
To COULTERVILLE
132
To TURLOCK LAKE
STANISLAUS COUNTY
MARIPOSA COUNTY
Lake McClure
To TURLOCK
Merced Falls
Yuba Con.
Snelling
San Joaquin
Hornitos
Lake Yosemite
59
MERCED COUNTY
0 1 2 3 4 5 6 7 8
To MARIPOSA
99
Merced
140

NEVADA COUNTY

Badger Hill

Location. This placer-mining district is in north-central Nevada County. The North Columbia district adjoins it on the east and the North San Juan district on the west. It was hydraulicked extensively from the 1850s through the 1880s. Later, Chinese miners reworked the tailings. It includes the Cherokee diggings. This district was recently studied by the U.S. Bureau of Mines and U.S. Geological Survey as part of their heavy metals programs.

Geology. The deposits are part of the Tertiary North Bloomfield-North Columbia-North San Juan channel. The gravels are thick, the lower part being quartzitic and the upper part containing abundant sand and clay. The lower gravels were rich. The upper gravels yielded 10 to 15 cents per yard. Bedrock is slate and phyllite, with granodiorite lying to the west. The value of the total output for the district is unknown, but it has been estimated at several million dollars. In 1891 the U.S. Army Engineers estimated that 10 million yards had been excavated and 33 million remained.

Blue Tent

Blue Tent is northeast of the Nevada City district in western Nevada County. The area was mined years ago by hydraulicking, but little or no mining has been done since. The gravels are part of the Tertiary channel that extends north-northwest from You Bet through Scotts Flat and Quaker Hill to North Columbia. Although the gravels here are extensive, they were reported not to have been very remunerative. Lindgren (1911) estimated that 15 million cubic yards had been removed and 90 million yards remain, much of it barren clay and sand. The gravel next to bedrock was reported to have yielded about 50 cents in gold per yard. To the east and south the gravels are overlain by andesite. Bedrock is phyllite and slate.

Emigrant Gap

Location and History. This district is in the east-central Placer and Nevada Counties in the vicinity of the towns of Emigrant Gap and Blue Canyon. It is both a lode and placer gold-mining district that was first worked during the gold rush. There has been intermittent prospecting and development work here since. The Zeibright mine was worked on a large scale during the 1930s. The Washington district is just to the northwest and the Westville district is to the south. The Zeibright mine camp is now a boy scout camp.

Geology. The district is chiefly underlain by thick beds of slate, schist and phyllite of the Blue Canyon Formation (Carboniferous). To the north and west these rocks are overlain by andesite and rhyolite tuff. There are several patches of auriferous Tertiary channel gravels.

Ore Deposits. There are a number of north-trending quartz veins in schist and slate that contain free gold, pyrite, and small amounts of other sulfides. The veins range from one to 10 feet in thickness and consist of a series of parallel quartz stringers. The milling grade ore usually contained 1/2 ounce per ton or less, but some of the ore shoots were extensive. Channel deposits at the Lost Camp hydraulic mine contain both cemented and free-washing quartzitic gravels. These gravels are overlain by sands and volcanic rocks. Some of the gold recovered from the gravels was coarse.

Mines. Placer: Golden Channel, Golden Nugget, Lost Camp, Shell, Wild Yankee.

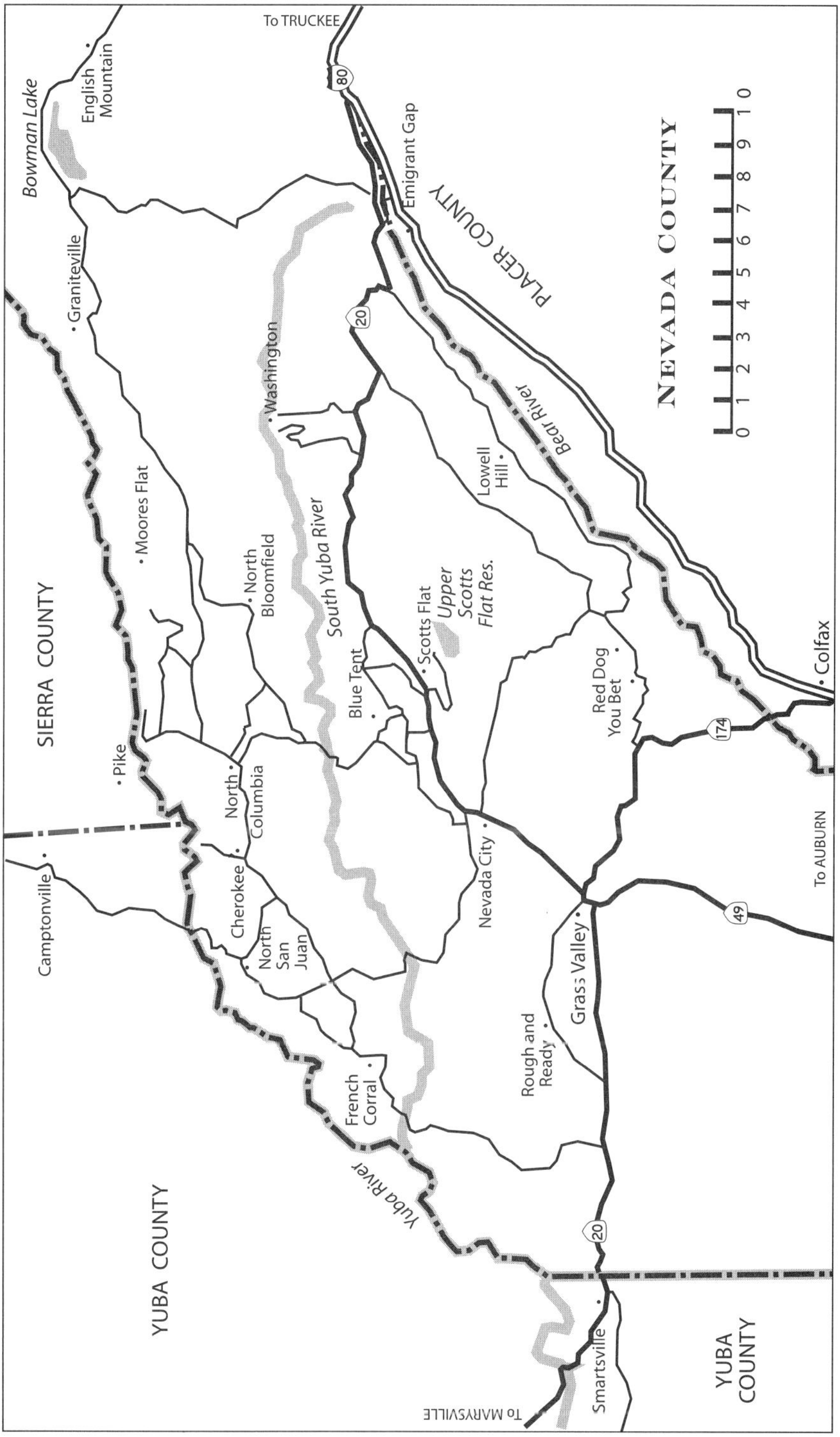
NEVADA COUNTY
0 1 2 3 4 5 6 7 8 9 10
To TRUCKEE
80
Bowman Lake
English Mountain
Emigrant Gap
PLACER COUNTY
Graniteville
20
Washington
Bear River
Lowell Hill
Moores Flat
South Yuba River
North Bloomfield
Upper Scotts Flat Res.
Scotts Flat
SIERRA COUNTY
Blue Tent
Red Dog
You Bet
Colfax
174
Pike
North Columbia
To AUBURN
Nevada City
Camptonville
Cherokee
North San Juan
49
Grass Valley
Rough and Ready
French Corral
Yuba River
YUBA COUNTY
20
Smartsville
YUBA COUNTY
To MARYSVILLE

English Mountain

Location. This district is in northeastern Nevada County in the vicinity of English Mountain and Bowman Lake. The area was first mined during the 1860s, followed by activity at the English Mountain gold and copper mines during the 1890s and early 1900s. There has been minor prospecting since.

Geology. The area is underlain by slate, schist, hornfels and granite, all of which are cut by diorite and quartz-diorite dikes. The ore deposits consist of iron-rich zones containing quartz, free gold, galena, pyrite, and chalcopyrite. Molybdenite is abundant in places. The principal gold sources have been the English Mountain Gold and Yellow Metal mines.

French Corral

Location. French Corral is in northwestern Nevada County about nine miles northwest of Nevada City. Much of the gold production came from hydraulic mines between here and Birchville to the northeast.

History. The town was named for a mule corral erected by a Frenchman, who was the first settler in the area. The principal period of gold mining was from the middle 1850s to the 1890s; there has been minor work since. Sometime before 1867, a 7-1/2-carat diamond, the largest known to have come from California, was found here in a sluice box.

French Corral was the terminus of one of the first long-distance telephone lines in the United States. Installed by the Edison Company about 1878, it connected Birchville, North San Juan and North Bloomfield to Bowman or French Lake, in the high Sierra Nevada some 58 miles east. It was used primarily to send messages about the delivery of water to the hydraulic mines, but it also was used by Western Union to send other messages.

The total production of the district is unknown, but it has been estimated to be valued at between $3 million and $4 million. Lindgren, in 1911, estimated that 32.5 million yards of gravel had been removed and that 20 million remained; the U.S. Army Corps of Engineers, in 1891, estimated that the same amount had been removed, but that only 10 million yards remained.

Geology. A major channel of the Tertiary Yuba River entered the area from the northeast. It extends southwest for a distance of about four miles in this district. The gravels deposited by this channel are 150 to 250 feet thick and 600 or more feet wide. The gravels have yielded gold throughout, but the quartz-rich lower gravels were the richest. Bedrock is granodiorite with greenstone to the north. Also there are some gold-quartz veins and bodies of mineralized granodiorite and greenstone.

Graniteville

Location and History. This district is in east-central Nevada County about 30 miles east of Nevada City. It is also known as the Eureka district. An extensive belt of gold mineralization in this region extends from the vicinity of the town of Graniteville south-southeast to the Emigrant Gap district, a distance of about 10 miles. The district includes the Gaston mine area, which sometimes has been classified as a separate district. The Allegheny-Washington gold belt lies a few miles to the west, and the American Hill district is to the north. The Graniteville district was first placer-mined during the gold rush, and quartz mining began soon afterward. Considerable mining activity continued from the 1860s until about 1900, and there was much activity again during the 1930s. The Ancho-Erie and a few other mines were worked for a short time after World War II.

Geology. The district is underlain by slate, schist, and phyllite of the Blue Canyon Formation (Carboniferous) in the west and granodiorite in the east. In addition, there are several patches of Tertiary gravel and several glacial moraines.

Ore Deposits. Three main north-striking vein systems run through the district. One in the western portion of the district is in slate and schist and contains the Culbertson, National, and Ancho-Erie mines. One to the east is in granodiorite and contains the Wisconsin, Baltic, and Iowa mines. The veins in the central system are along the slate-granodiorite contact. Properties in the central system include the Rocky Glen and Gaston mines. The quartz veins are as much as 15 feet thick. The ore contains free gold and varying amounts of auriferous sulfides. The milling ore usually averages 1/2 ounce per ton or less with very little high grade. Some of the ore shoots had stoping lengths of several hundred feet. The veins were mined to depths of as much as 500 feet. The Tertiary gravels at Graniteville and at Shands two miles to the west have yielded some gold.

Mines. Alpha, Ancho-Erie $1 million+, Anderson, Attic, Azalie, Baltic, Barren, Birchville, Blue Bell, Celina Flat, Cooley, Culbertson, Eagle Bird, Gaston $2 million?, German, Gold Bug, Hotwater, Iowa, IXL, Jim, Keller, Last Chance, Lindsay, Mountain View, National, Rainbow, Rattlesnake, Republic, Rocky Glen $300,000+, Star, Washington, Wisconsin, Yellow Metal, Yuba $2 million+.

Grass Valley

Location. This famous mining district is in western Nevada County in the immediate area of the town of Grass Valley. The Nevada City district adjoins it on the northeast and the Rough-and-Ready district is to the west.

History. Placer gold was first found in Wolf Creek in 1848 shortly after Marshall's discovery at Coloma. The earliest mining was done by David Stump and two companions who came from Oregon. The shallow placers were rich but were exhausted quickly. Gold bearing quartz was discovered at Gold Hill in 1850 and soon afterward at Ophir, Rich, and Massachusetts Hills. Quartz mining soon developed into a major industry that was to last more than 100 years. The Gold Hill and Allison Ranch were the leading lode mines during the 1850s. Mining was curtailed somewhat during the Comstock rush of 1859-65, but the mines were productive again in the late 1860s. The camp declined in the 1870s, and by 1880 only the Empire and Idaho mines were active. In 1884 the North Star mine was reopened and activities increased; the North Star, Empire, Idaho-Maryland, Pennsylvania, and W.Y.O.D. all were highly productive. By 1900, the Idaho-Maryland mine had yielded a total of $12.5 million. From 1900 to 1925, the North Star and Empire mines were the largest producers, the Idaho-Maryland having been idle in 1901-19. By 1928, the North Star had had a total output valued at $33 million.

In 1929 the Empire and North Star groups were purchased by the Newmont Mining Corporation. This merger, which resulted in the Empire-Star Mines Co., included other important mines, such as the Pennsylvania, W.Y.O.D., and Sultana. From 1930 to 1941, the district was enormously productive. The 1930-40 output of Idaho-Maryland Mines Corp., which included the Idaho-Maryland and Brunswick mines, was $26.7 million. The Empire-Star group yielded 1,074,284 ounces of gold from 1929 to 1940. Nearly 4000 miners were at work in the mines during the 1930s and early 1940s. The mines were shut down during World War II, but the Empire, Pennsylvania, North Star, and Idaho-Maryland reopened soon afterward. However, operations gradually decreased; the Idaho-Maryland stopped gold mining in 1956 and the Empire-Star group in 1957, closures that ended nearly 106 years of gold-mining operations in the Grass Valley district. Some tungsten ore was mined in 1954-57 at the New Brunswick unit of the Idaho-Maryland mine.

Grass Valley was the richest and most famous gold mining district in California. The value of the total output of the lode mines is estimated to have been at least $300 million, and the placer mines yielded a few million dollars more worth of gold. The two largest operations, the Empire-Star and Idaho-Maryland groups, had total outputs of $130 million and $70 million, respectively. Many famous mining engineers and geologists worked in the Grass Valley district. A number of important inventions and improvements were made in mining and milling equipment in the Grass Valley gold mines. Many of the miners were of Cornish descent and were often known as "Cousin Jacks". For many years the town and the mines were served by the Nevada County Narrow Gauge Railroad, which extended north from Colfax. A few historic mine structures are still standing, but most of the extensive surface plants of the major mines have been dismantled. The old power house at the North Star mine and its 32-foot Pelton wheel are part of a Nevada County historical display.

Geology. An elongated body of granodiorite is in the central portion of the district. This body is five miles long in a north-south direction and 1/2 to two miles wide. It is intrusive into older metamorphic rocks and itself is cut by various dike rocks. Immediately east and west of the intrusion are dark greenstones classified as metadiabase and metadiabase porphyry (so-called "porphyrites"), and continuing to the northeast are amphibolite schist, serpentine, gabbro and diorite, and slate. Just north of the granodiorite and to the southwest are slates, phyllite, quartzite, and schist of the Calaveras Formation (Carboniferous to Permian). A number of intermediate to basic dikes are present also, as well as a few aplite and granite porphyry dikes. Overlying part of the district to the east and to the northwest are Tertiary gravels, in turn largely overlain by andesite.

Ore Deposits. This is the most heavily mineralized and richest gold district in the state with a very large number of productive veins in a relatively small area. The veins fall into two major groups: 1) those of the granodiorite-greenstone area, which have gentle dips, and 2) those of the serpentine-amphibolite area, with steep dips. The veins of the granodiorite area are either in the granodiorite or in the adjacent greenstone, entering the granodiorite at depth. One group of veins strikes north and dips gently (about 35' on the average) either east or west. This group includes the Empire, Pennsylvania, Osborne Hill, Omaha, W.Y.O.D., and Allison Ranch veins. The other group of veins in the granodiorite strikes west or northwest and dips gently north. The North Star and New York Hill veins are included in this group. In the serpentine-amphibolite area the veins strike northwest and dip steeply southwest; a few dip northeast. These occur mostly in the amphibolite near or at the serpentine contact. The Idaho-Maryland, Brunswick, and Union Hill mines are here.

The veins usually range from one to 10 feet in thickness and consist of quartz with some calcite and ankerite. They fill minor thrust faults. Many veins contain several generations of quartz. There are numerous northeast-striking, vertical or steeply-dipping fractures or "crossings" that commonly are boundaries of ore shoots. The ore contains free gold and varying amounts of sulfides, chiefly pyrite. Present in smaller amounts are galena, chalcopyrite, arsenopyrite, sphalerite, and pyrrhotite. Galena is commonly associated with gold.

The ore shoots vary considerably in size and shape, and the distribution of gold within the shoots is erratic. Some have pitch lengths of up to several thousand feet, and the veins have been developed to inclined depths of as much as 11,000 feet. Much specimen ore has been found, but milling ore usually averaged from 0.25 to 0.5 ounce of gold per ton. Coarse-grained scheelite is present in several veins, notably in the Union Hill and New Brunswick mines.

Mines. Ajax, Alaska, Alcade, Allison Ranch $2.7 million, Alpha, Bella Union, Ben Franklin $750,000, Big Diamond, Black Hawk, Bow, Boundary, Buena Vista, Bullion, Cassidy (Linden), Centennial $500,000+, Cheranne, Coe $500,000+, Conlan, Crown Point * $130,000+, Daisy Hill , Dakota *, Diamond, East Star *, Empire *, Empire-Star group $130 million, Empire West 1, Empress, Eureka $5.7 million, Gaston, General Grant, Gladstone, Golden Center (Dromedary) $2.5 million+, Golden Gate, Golden Treasure, Gold Hill *, Gold Point, Goodall, Granite Hill, Grant, Hartery $350,000, Hermosa, Houston, Homeward Bound , Houston Hill, Idaho, Idaho-Maryland group $70 million, Independence, Inkmarque, Kate Hayes *, Larimer, Le Duc, Lone Jack, Magenta, Mary Ann, Maryland, Massachusetts, Massachusetts Hill *, New Brunswick, New Eureka, New Homeward Bound, New Ophir *, New York Hill, Normandy-Dulmaine, Northern Bell, North Star *, Norumbagua $1 million+, Oakland $100,000, Old Brunswick, Old Eureka, Old Homeward Bound, Omaha, Orleans, Osborne Hill, Peabody, Pennsylvania, Phoenix, Polar Star, Prescott Hill, Prudential $100,000, Republic, Reward, Rich Hill, Rocky Bar, Rose Hill $100,000+, Scotia, Sebastopol, South Idaho, Spring Hill $300,000+, Stockton Hill, St. John, Sultana *, Syndicate, Telegraph*, Union Hill $750,000, Wisconsin, Wyoming, W.Y.O.D.

Lowell Hill

Location and History. This district is in south-central Nevada County about six miles northeast of Dutch Flat. It includes the Remington Hill, Negro Jack Hill, and Liberty Hill areas. The district was hydraulicked from the middle 1850s through the 1870s, and Liberty Hill was worked again from around 1896 to 1915.

The total output is unknown, but it exceeds $1 million. Lindgren, in 1911, estimated that two million yards had been removed and 16 million remained at Liberty Hill and that 1.75 million yards had been removed and 6 million remained at Remington Hill. Other estimates of remaining gravel at Liberty Hill range from six to 10 million yards.

Geology. The deposits are in a southwest-trending Tertiary channel that joins the Dutch Flat channel. There is a lower well-cemented blue gravel that contains gabbro and serpentine boulders and yielded 18 to 23 cents a yard. An upper quartz-rich gravel is in places covered by heavy clay. Bedrock is slate of the Blue Canyon Formation (Carboniferous) and some serpentine.

Moore's Flat

Location. The Moore's Flat district is in north-central Nevada County about 15 miles northeast of Nevada City. It is both a lode and placer district and includes the "diggings" at Moore's Flat, Oreleans Flat, Woolsey Flat, Snow Point, and Snow Tent. The Allegheny district adjoins the Moore's Flat district on the northeast. Moore's Flat was named for H. M. Moore who built a store there in 1851.

Geology. A number of gravel deposits were accumulated in a west-southwest-trending Tertiary channel of the Yuba River that continues west and southwest into the North Bloomfield district. At Moore's Flat, Lindgren (1911, p. 141) estimated, 26 million cubic yards were removed and 15 million remained. The gravels are quartz-rich and in places more than 100 feet thick. Hydraulic mining here during the 1880s had reported gold recoveries of 11 to 15 cents per yard. The gravels are capped by andesite on the south side of the district. Bedrock consists of amphibolite, slate, and serpentine. The gold-quartz veins usually are narrow and contain small but often rich pockets.

Nevada City

Location. The Nevada City district is in western Nevada County. The district covers an extensive area, from the vicinity of Indian Flat east through Nevada City, northeast to Willow Valley and southeast through Canada Hill and Banner Hill to the vicinity of the Lava Cap mine. It is both a lode- and placer-mining district and once was an important center of gold mining in California.

History. Gold was first mined in this district in Deer Creek, which, in 1849, was called Deer Creek Diggings. The name Nevada was adopted in May 1850 at a public meeting. The placers were rich, and the town grew fast. Hydraulic mining was first practiced in California at American Hill here in 1852, by E. G. Matteson (hydraulicking was also done that same year at Yankee Jim's in Placer County). Hydraulic mining flourished until around 1880. Drift mining began in the 1850s, and the drift mines were continuously active until around 1900. Gold-quartz was discovered in 1850, when the Gold Tunnel vein was found. However, important production of lode gold did not commence until the early 1860s because of difficulties in milling the ore. By 1865, the output from lode-gold mining was averaging $500,000 per year and later ranged from $300,000 to $600,000. The Champion and Providence mines were the major producers during these years. Later these two mines were in litigation, and in 1902 the Champion owners bought the Providence. Large-scale lode-gold mining was resumed in the district again during the 1930s when the Lava Cap and Banner mines were operating. From 1933 to 1942, the Lava Cap yielded $12 million. There has been only minor activity since 1942. The old town of Nevada City, the county seat, is now a popular tourist center with numerous well-preserved old buildings. The Nevada County Narrow Gauge railroad served the area from 1877 to 1942. The total output of the district is unknown, but it is estimated by the author at more than $50 million and may have exceeded $70 million.

Geology. The central portion of the district is underlain by granitic rocks, chiefly granodiorite. Adjacent are beds of slate, mica schist, and quartzite, most of which are part of the Calaveras Formation (Carboniferous to Permian). To the west and southwest are fairly extensive beds of massive greenstone, amphibolite, and serpentine. There are a number of fine- to medium-grained dioritic and aplitic dikes, some of which are associated with the gold-quartz veins. In places these rocks are overlain by Tertiary channel gravels capped by rhyolite and andesite.

Ore Deposits. Several major gold-quartz vein systems traverse the district. In the west portion one system extends northwest along a granodiorite-metasedimentary rock contact. In the southern and eastern portion of the district the veins strike nearly west and dip either north or south. There are also a few northeast-striking and southeast-dipping veins. The veins usually are one to four feet thick, but in places a few are as much as 15 feet thick. The ore contains varying amounts of free gold, often abundant pyrite and smaller amounts of other sulfides. Some of the ore bodies are extensive; the ore body at the Providence mine persisted to an inclined depth of more than 2700 feet. Considerable high-grade ore has been recovered in the district.

Several important Tertiary channels were sources of ore-bearing gravels. One, the Harmony channel, which enters the district from the northeast, was extensively mined by drifting.

The pay gravel in this channel was 150 to 200 feet wide, two to four feet deep, quartzitic, often sub-angular and well-cemented. These pay streaks yielded $1.55 to $2.50 in gold to the ton at the old price. The Manzanita channel, which yielded $3 million, is just to the west. Northwest of town is the northwest-trending Cement Hill channel. In the southern part of the district is the Town Talk channel, which was narrow but rich in places. Much of the placer gold taken from the channel deposits in this district was coarse.

Mines. Hydraulic: American Hill, Buckeye Hill, Canada Hill, Hirschmann.

North Bloomfield

Location. The North Bloomfield mining district is in north-central Nevada County about 10 miles northeast of Nevada City. This district also includes the "diggings" at Lake City to the west, Derbec to the north, and Relief to the east.

History. Gold was discovered here originally in 1851. Hydraulic mining began about 1853 and, by 1855, had become a major industry. An extensive system of ditches and flumes supplied water to the mines from Bowman Lake and other reservoirs to the east in the high Sierra Nevada. The town of North Bloomfield was first known as Humbug City. Its name was changed to Bloomfield and then to North Bloomfield when someone discovered there was a Bloomfield in Sonoma County.

As more and more gold-bearing gravel was excavated, the hydraulic pits here became enormous. The pit at the famous Malakoff mine is more than 7000 feet long, 3000 feet wide, and up to 600 feet deep. The tailings from the hydraulic operations were allowed to flow into the rivers, a procedure that led to litigation with the farmers who lived downstream. In a famous court case in 1884 (Woodruff vs. North Bloomfield Gravel Mining Company [16 Fed. Rep. 25]), judge Lorenzo Sawyer issued an injunction against the dumping of mine debris into the Sacramento and San Joaquin Rivers and their tributaries. Injunctions against other mines soon followed, and hydraulic mining in the Sierra Nevada has not been important since that date. The Malakoff diggings and part of the old town are now a state park. The U.S. Geological Survey, beginning in 1966, initiated an exploration program in this area to determine the extent of unmined gravels. Drilling and geophysical exploration have been done.

The total output of the Malakoff mine is about $3.5 million, and the Derbec mine has probably yielded $l million to $2 million according to Lindgren (1911). He estimated that 30 million yards had been removed and 130 million remained at North Bloomfield; Jarman (1927) estimated that 40 million yards had been removed and more than 50 million remained.

Geology. The main channel of the Tertiary Yuba River entered this district from the northeast via Derbec. A branch joined this channel from Relief Hill to the southeast. At North Bloomfield the main channel curves west and north and then west again as it continued toward the North Columbia district. Although the gravels are as much as 600 feet thick, most of the values were obtained from the lower 130 feet of blue gravel. These gravels yielded from four to 10 cents of gold per cubic yard. Bedrock consists of slate, schist, and phyllite.

North Columbia

Location. This district is in north-central Nevada County about seven miles northeast of Nevada City. It includes placer deposits in the North Columbia and Columbia Hill areas and lode deposits in the Delhi mine area.

History. The district was first placer-mined during the gold rush and hydraulicked on a large scale from the middle 1850s to the early 1880s. It was named after Columbia Hill, and "North" was added to distinguish it from Columbia in Tuolumne County. The Delhi lode mine was active from the 1860s to the 1890s. Chinese carried on small-scale placer mining from the 1890s to the early 1900s. Some work was done again during the 1930s and the area has been prospected since. The total output of the district is unknown. The placer mines are estimated to have yielded $2 million to $3 million. Lindgren (1911) estimated that 25 million yards of gravel had been removed and 165 million remained. The lode mines have yielded more than $1 million. Beginning in 1966, the U.S. Geological Survey and U.S. Bureau of Mines have studied this district as part of their "heavy metals" programs (see also sections on Badger Hill, and North Bloomfield).

Geology. Extensive channel deposits lie at the junction of two major streams of the Tertiary Yuba River. A west-trending channel extends through the district from North Bloomfield and continues on to Badger Hill, and a branch enters the district from Blue Tent to the south. In places the gravels are as much as 500 feet thick. The lower gravels are coarse while upper bench gravels are fine. Much sand and clay are present. Bedrock is phyllite of the Delhi formation (Carboniferous). Several gold-quartz veins have yielded high-grade ore.

Mines (all lode). Delhi $1 million, Enterprise, Grizzly Ridge, St. Gerhard.

North San Juan

Location. This district is in northwestern Nevada County nine miles northwest of Nevada City. It consists of the placer deposits that extend from the vicinity of North San Juan southwest for about 2-1/2 miles to the Sweetland area. It was named for a nearby hill with the "North" added to distinguish it from San Juan (now San Juan Bautista) in San Benito County.

Geology. A main Tertiary channel of the Yuba River extends into the area from the Badger Hill and North Columbia districts from the east and another from Camptonville from the northwest. The channel then extends southwest through Sweetland into the French Corral district. The gravels are 150 to 400 feet thick and up to 1000 feet wide. The low gravels yielded up to 300 per yard in gold, mostly from hydraulicking. The bedrock is granodiorite with amphibolite to the west. There are several narrow gold-quartz veins in the amphibolite.

Rough-and-Ready

Location and History. Rough-and-Ready is in western Nevada County about five miles west of Grass Valley. Placer mining began here during the gold rush. The town was founded in 1849 by the Rough and Ready military company led by Captain A. A. Townsend. He had once served under General Zachary Taylor, who was known as "Old Rough and Ready". Drift and hydraulic mining began in the late 1850s, and lode mining became important in the 1860s. There was considerable activity that lasted until about 1900. Some work was done in the district in the 1930s, and there has been minor prospecting since.

Geology and Ore Deposits. A north-trending belt of amphibolite one to two miles wide traverses the central portion of the district. Gabbro and diorite lie to the east and granodiorite to the north. Some orbicular gabbro is present. A west-trending Tertiary gravel channel of the Yuba River crosses the north portion of the district. A number of north-striking quartz veins occur chiefly in amphibolite. The veins are one to five feet thick and contain free gold with pyrite and other sulfides. Considerable high-grade ore was taken out. None of the veins has been developed to depths of more than 500 feet.

Mines. Placer: Alta-California, Jenny Lind.

Scotts Flat

Location and History. This district is in west-central Nevada County about seven miles due east of Nevada City. It includes the Tertiary placer "diggings" at Scotts Flat, Quaker Hill, Hunts Hill, Buckeye Hill, and Burrington Hill. The You Bet-Red Dog district lies immediately to the south and the Nevada City district to the west. The various mines were extensively hydraulicked from the 1850s through the 1880s, and later the tailings were reworked by Chinese miners. Also there was drift mining in the district. The area was prospected during the 1930s.

Geology. These deposits are in the north-northwest-trending Tertiary gravel channel that extends from You Bet-Red Dog to North Columbia. A southwest-trending tributary comes into the area from Burrington Hill and joins this channel at Hunts Hill. At Hunts Hill and Quaker Hill the main channel is nearly 600 feet deep with bench gravels up to 300 feet in depth. The deep gravels are well-cemented and quartz-rich and, in places, were very rich in gold. The upper gravels usually are fine and contain abundant sand. The deep channel is believed to be continuous all the way from Hunts Hill to the Blue Tent district, a distance of seven miles. Bedrock in the east portion is slate and in the west, phyllite and greenstone. On the major ridges the gravels are capped by Tertiary andesite and rhyolite.

In 1911 Lindgren estimated that 12 million cubic yards of gravel had been removed from Scotts Flat and 35 million from Quaker Hill. He also estimated that a vast amount (140 million cubic yards) remained at Quaker Hill. The U.S. Army Engineers (Jarman, 1927) estimated 50 million to 90 million cubic yards remained at Quaker Hill. They also estimated that 6.75 million cubic yards had been removed, and 4 million to 5 million remained at Hunts Hill.

Smartsville

See text under Yuba County, p. 183.

Washington

Location. This district is in east-central Nevada County in the vicinity of the old mining town of Washington, 18 miles northeast of Nevada City. It is in the south end of the Goodyear's Bar-Allegheny belt and includes the "diggings" at Alpha and Omega.

History. The Washington district was first mined during the gold rush, and the placers of the Middle Yuba River were highly productive. The Omega and Alpha hydraulic mines were opened in the middle 1850s and worked on a major scale through the 1880s. Later, Chinese miners reworked the tailings. Lode mining also began in the 1850s and continued steadily until about 1915. There was activity in the district again during the 1930s, and the Red Ledge mine has been worked in recent years. Barite, chromite, and asbestos also have been mined here.

Geology. The district is underlain chiefly by slate, schist and quartzite of the Blue Canyon Formation (Carboniferous). A serpentine body one to two miles wide crops out in the central portion. The Relief Quartzite (Carboniferous) and amphibolite lie to the west and granodiorite to the east. The serpentine is a south extension of a belt that passes north-northwest through Allegheny and Goodyear's Bar in Sierra County. Tertiary andesite overlies the main ridges to the north and south.

Ore Deposits. The auriferous Tertiary channel gravels at Alpha and Omega are part of the main channel that extends west and north to Relief and North Bloomfield. Jarman (1927) estimated that, at Omega, 13 million yards were mined and yielded 13-1/2, cents in gold per yard. Lindgren (1911) estimated that 40 million yards remained. The quartz veins contain small but rich ore bodies, similar to those of the Allegheny district to the north, but are not as plentiful. Except for arsenopyrite, sulfides are not usually abundant. A number of beautiful specimens of crystallized gold have been found in the Red Ledge mine. The Spanish mine also has yielded large amounts of barite.

Mines. Placer: Alpha $2 million+, Centennial, Omega, Phelps, Yuba River.

You Bet

Location and History. You Bet and Red Dog are in south-central Nevada County, eight miles southeast of Nevada City. This district also includes the "diggings" at Little York. You Bet sometimes is known as Chalk Bluffs. The region was first placer-mined in 1848 or 1849. The name "You Bet" is supposed to have originated in 1857 from saloon keeper Lazarus Beard's favorite expression. Red Dog was named by Charley Wilson after his former home, Red Dog Hill, Illinois. The district was hydraulicked on a large scale from 1855 until the 1880s. There was some drift mining. Later, the area was mined on a moderate scale, chiefly by Chinese. It was intermittently active until about 1935. The total output is valued at more than $3 million. Lindgren (1911) estimated 47 million yards were removed and 100 million remained. Jarman (1927) estimated 20 million yards of 10- to 15-cent gravel remained at Red Dog.

Geology. The gravels were deposited by a Tertiary channel of the Yuba River that extended north and northwest to Hunts Hill and Scotts Flat. The lower gravels are 30 to 40 feet thick, well-cemented, and contain a high percentage of quartz including a number of large boulders. It is capped by as much as 350 feet of fine gravel with some interstratified clay and sand. Bedrock consists of slate and some chert.

• Nevada County Placer Operations

Calaveras Gold Dredging Company operated a dragline dredge at Steep Hollow in the You Bet district from April 8 to August 20, 1940.

Clerkin Property consists of 156 acres of patented land in sec. 24, T. 17 N., R. 7 E., near French Corral, owned by W. P. Clerkin. Clerkin states the ground contains 30 to 40 acres of gold-bearing gravel, of which a 60- to 70-foot depth contains practically no gold. From 8 to 10 feet at the bottom is blue gravel, and the last foot contains most of the gold. Clerkin thinks that this deposit represents a high bench of the old Eocene channel about 100 feet wide and adjoins the bedrock tunnel claim of the French Corral placer.

Dakin Company, San Francisco, operated a dragline dredge at Champion Flat along Deer Creek from January 1 to March 1, 1940.

French Corral Placer is a property of 1,700 acres in secs. 23, 24, 25, 35, T. 17 N., R. 7 E., including the Kate Hayes, Esperance, Fraser, and Alexander, Bedrock Tunnel, and Milton placer mines. The property contains approximately 1 mile of unworked length of the Eocene channel worked by hydraulic methods at North Bloomfield, North Columbia, Badger Hill, American Hill, North San Juan, Sweetland, Birchville, and French Corral.

Greenhorn Dredging Company operated a dragline dredge at Quaker Hill in the You Bet district from January 1 to December 31, 1940.

Omega mine is in Sec. 16, 17, T. 17 N., R. 11 E., 3 miles southeast of the town of Washington. The last mining was done in the season 1941-42. During the season the company washed 429,637 cubic yards of gravel, which yielded 1302 ounces of gold and 49 ounces of silver.

William Richter & Sons operated a dragline dredge at Scotts Flat from January 1 to October 13, 1940.

San Juan Gold Company, San Francisco, holds 5484 acres of patented mining claims on San Juan Ridge principally in the following sections: Sec. 36, T. 18 N., R. 8 E.; Sec. 1, T. 17 N., R. 8 E.; Secs. 1, 2, 4, 5, 6, 7, 8, 9, 11, 12, T. 17 N., R. 9 E.; Secs. 35, 36, T. 18 N., R. 9 E.; Secs. 16, 21, 22, 31, T. 18 N., R. 10 E. These holdings include the famous Malakoff hydraulic mine.

Western Gold Incorporated operated the Relief Hill mine by the hydraulic method in 1941 and was one of the first companies to utilize debris storage behind the Upper Narrows Dam on the Yuba River. The mine is 3 miles east of North Bloomfield in secs. 4, 9, T. 17 N., R. 10 E. Drift mining has been done at this property as well as hydraulic mining.

You Bet mines include many of the old hydraulic mines famous in the early days of that method of mining. The group includes 1150 acres of patented land and 120 acres of locations in secs. 25, 26, 36, T. 16 N., R. 9 E.; secs. 29, 30, 31, 32, T. 16 N., R. 10 E.; secs. 2, 11, 14, 15, T. 15 N., R. 9 E.; sec. 6, T. 15 N., R. 10 E. Immense yardages of gravel were washed at these properties by the hydraulic method in the early days from one of the great Tertiary river channels, but large yardages still remain unworked. The Jarman report of 1927 estimates that the property still contains 12,000,000 to 24,000,000 cubic yards of gravel containing enough gold to pay for hydraulic mining. Tom Brady, who holds the adjoining Jupiter group of 100 acres in sec. 31, T. 16 N., R. 10 E., states that a part of the channel starting at a point east of You Bet and running northward has never been worked, except a little on the surface. [1]

PLACER COUNTY

• TERTIARY CHANNELS

Eocene Channels. Lindgren (1911) has suggested one main channel to include most of the big patches of auriferous gravels in the eastern half of the area. According to him, this channel enters the area at Michigan Bluff, passes near Sage Hill, runs west to the Paragon Mine, then north and west to the Mayflower Mine and then south to the San Francisco Mine. From here on, this channel takes a less discernible course to Yankee Jim, Smith's Point, then to King's Hill, Big Dipper Mine, Morning Star Mine, Iowa Hill, and finally north to Gold Run.

From Yankee Jim on, this channel travels somewhat upgrade in its flow to Dutch Flat. He attributed the apparent upward flow to the westward tilting of the Sierran block. In his opinion, as the stream was flowing south to north and not south-southeast to north-northwest, the west-southwestward tilting of the Sierran block caused the present-day upgrade of the supposed course of the lone Channel, that is, this tilting produced higher bedrock elevations at Big Dipper Mine, California Morning Star Mine, and Iowa Hill successively from the almost level course of the stream.

Regarding the course of the channel, the writer agrees with Lindgren as far as the site of the Foresthill placer mines. From the Mayflower Mine onward, however, the channel more plausibly ran south to Baltimore, thence southwest to the San Francisco Mine and Pond's Pit, and finally south to join with any stream flowing in the general direction of present flow of the Middle Fork of the American River. The elevation of the bedrock at Pond's Pit is not too great to have prevented the channel from taking that course.

A little west of this channel is another smaller channel, coming from the direction of Cottage Home Hill and extending through Buckeye Mine, Keystone Mine, Adam's Pit, Smith's Point, Small Hope Mine, Yankee Jim (where the channel forms a bar), thence to the placer diggings south of Devil's Canyon and finally towards Grey Eagle Mine.

The most important channel is the one which comes from Gold Run, enters the area at Roach Hill, thence crosses Indian Canyon to form the lucrative Iowa Hill Mine deposits, recrosses Indian Canyon to form the Iowa Hill deposits, and then passes through the California Morning Star Mine. Thus the channel follows a meandering course to the Morning Star Mine. From there, it flows east, then southwest to the Big Dipper diggings, where it receives a tributary coming from Grizzly Flat. From the Big Dipper Mine, the channel flowed southwest to join with a tributary coming along King's Hill and King's Hill Point. At Roach Hill the former channel joins another tributary extending from North Succor Flat via Penn Valley, and Monona Flat.

The channel along King's Hill probably drained the Second Sugar Loaf area, and flowed through King's Hill Point to join with the Roach Hill-Iowa Hill-Big Dipper Mine channel; and the joint channel might have run in the general downstream direction of the North Fork of the American River.

As the area is close to the foothills, the tilting due to the Sierran block movement was here probably less than 1°. This is evident from a study of the profile of the Eocene surface from west to east. Furthermore, a greater tilt would go against the possibility of flow of any stream from Michigan Bluff to Paragon Mine. Such an insignificant tilting would cause no appreciable difference in the relative elevation of the bedrocks of the two streams, one flowing

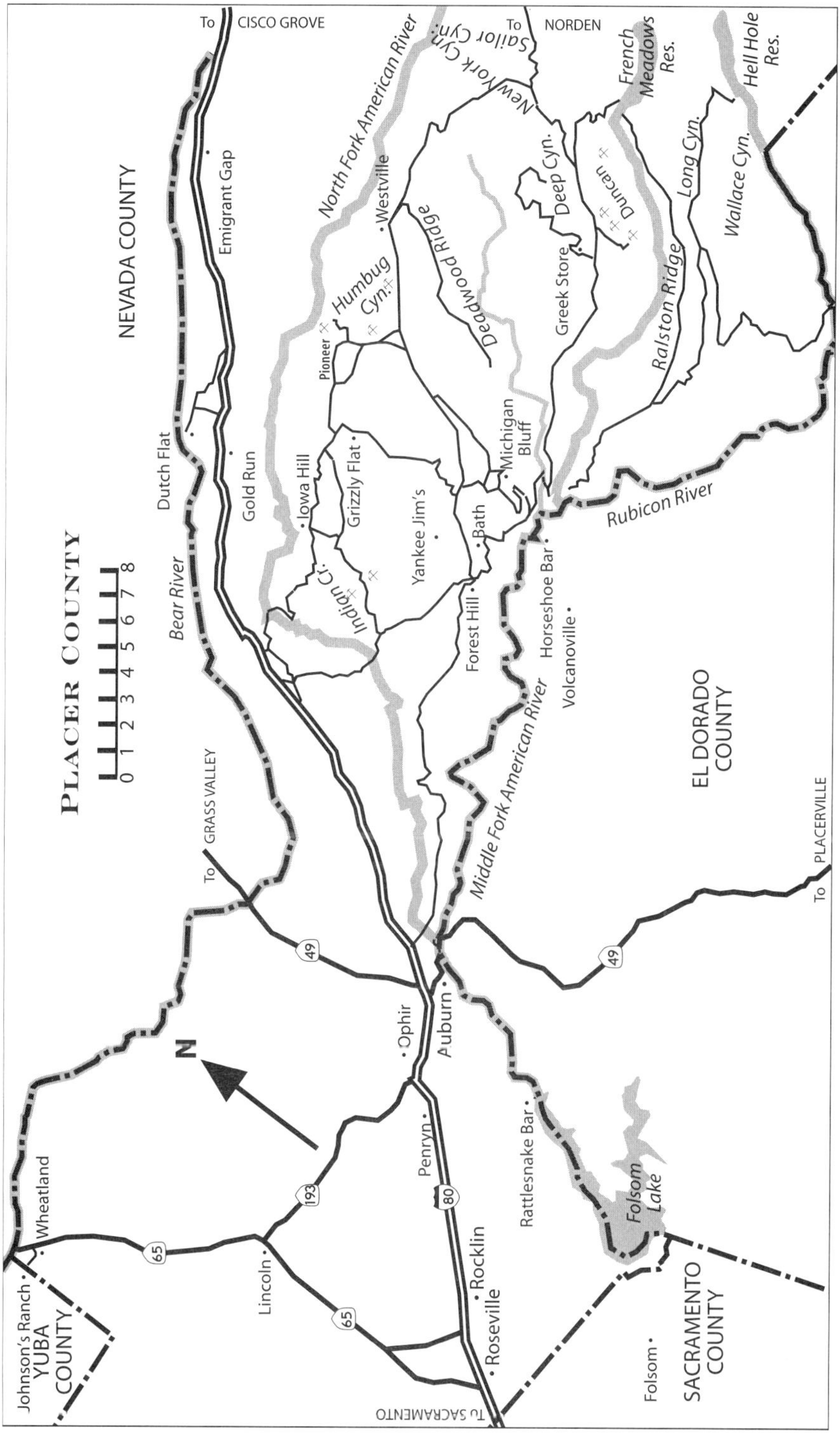
PLACER COUNTY
0 1 2 3 4 5 6 7 8
N
NEVADA COUNTY
EL DORADO COUNTY
SACRAMENTO COUNTY
YUBA COUNTY
To CISCO GROVE
To NORDEN
To GRASS VALLEY
To PLACERVILLE
To SACRAMENTO
North Fork American River
Middle Fork American River
Rubicon River
Bear River
Sailor Cyn.
New York Cyn.
French Meadows Res.
Hell Hole Res.
Long Cyn.
Wallace Cyn.
Deep Cyn.
Duncan
Ralston Ridge
Deadwood Ridge
Humbug Cyn.
Indian Cr.
Emigrant Gap
Westville
Pioneer
Greek Store
Michigan Bluff
Iowa Hill
Grizzly Flat
Yankee Jim's
Bath
Forest Hill
Horseshoe Bar
Volcanoville
Dutch Flat
Gold Run
Auburn
Ophir
Penryn
Rattlesnake Bar
Folsom Lake
Folsom
Wheatland
Johnson's Ranch
Lincoln
Rocklin
Roseville
49
65
193
80

north, the other north-northwest. The writer checked the elevation of bedrocks at the mines from the topographic maps and found that the bedrock elevation at Iowa Hill is higher than at California Morning Star Mine, while the latter is higher than the bedrock elevation at Big Dipper Mine. This suggests that the flow direction of the Eocene channel was from Iowa Hill to California Morning Star Mine and thence to Big Dipper Mine.

Lindgren did not include Roach Hill in this channel for the obvious reason that the elevation of the bedrock at Roach Hill was 300 feet higher than at Iowa Hill. The bedrock elevation at Gold Run is about 350 feet higher than at Roach Hill. This also suggests that the flow direction of the Eocene channel was from Gold Run to Roach Hill and thence to Iowa Hill.

Intervolcanic Channels. In most of the placer mines, "auriferous gravels" are represented by deep gravels. Bench gravels and other flood plain deposits belonging to the same formation are generally lacking except in a few places. In one such place, Elizabeth Hill Mine, bench gravels, sands, and carbonaceous shale are present. These are overlain by the Mehrten formation without any intervening Valley Springs formation, indicating no intervolcanic stream eroded the flood-plain deposits left by the Eocene stream. In other placer mines, the deep gravels are generally overlain by gravels of the Valley Springs formation, believed derived from the auriferous gravels and redeposited by the intervolcanic streams along with some rhyolite detritus. On this basis, this writer finds that for most of the Eocene channels there is a corresponding intervolcanic channel following almost the same course as the Eocene stream.

According to Lindgren, the lone channel from Foresthill onward was different from the intervolcanic channel which, in his opinion, agreed with present-day drainage. This writer suggests that the intervolcanic streams followed more or less the same courses as the Eocene streams, and that the main channels of both periods followed the general direction of present-day main drainage. It should be noted, however, that the drainage patterns of Eocene and intervolcanic channels were different from that of present-day channels, that is, the Eocene and intervolcanic streams, overloaded with sediments and flowing on a flat surface, formed meandering courses, whereas the present-day streams, being in the youthful stage, flow more or less straight.

ECONOMIC GEOLOGY

Gold Placer Mines

When the easily-won placer gold deposits along the banks and beds of the present-day rivers and streams were exhausted, search for pay gravels continued higher up in the hills. Thus surface gravels such as those at Iowa Hill, Yankee Jim's, were explored. This led to the search for the continuation of such channels below the thick cap of volcanic muds and boulders. In most places, two or three channels are superimposed, the older channel being generally richer in gold content than the younger ones. Thus the Eocene channels contain more placer gold than the intervolcanic ones which, moreover, derived most of their placer gold by reworking the gravels of the former channels. The major part of the placer gold found along the banks and beds of the present channels, was derived principally from the reworking of the auriferous gravels of the Eocene and intervolcanic channels.

Before hydraulic mining was introduced the placer deposits in the area were worked by drifting. When restrictions were imposed on the hydraulicking, drift mining once more became an important method of mining. In the placer deposits, the pay-streaks were usually on or near bedrock; and if there was a thick unproductive andesitic rhyolitic mantle, mining by hydraulicking became uneconomical, especially, when there was insufficient water or fall of water. In such cases, drift mining was the only feasible method. In several mines, however, where the mantle over the pay dirt was rich enough to pay for washing, hydraulicking successfully replaced drifting.

The drift mines in the area were worked by adits driven along or close to the trough of the channel, so that the pay dirt could be worked directly by the levels or raises. Where there was a thick overburden, shafts, usually vertical, but occasionally inclined, were sunk to or beyond the drift which ran along the course of the old channel. In order to prospect the whole width of the channel, crosscuts at right angles to the drift were driven on both sides. In a few mines, adits were driven in the competent bedrocks, just a few feet below the old channel bed, and the pay dirt lying in the channel bed was tapped by use of raises. By this method, mining was done without much timbering. The channel gravels in some mines, however, were cemented and were strong enough to stand without timbering. The drifts were driven in such a way as to allow drainage out of the entrance adit. Besides the main shaft for hoisting ores or pumping water, there were one or more vertical shafts for ventilation in mines, especially where the drifts extended for long distances.

The major placer deposits in the area were at one time or another mined by hydraulicking. In most cases water had to be brought in open ditches, pipes or flumes from long distances. These ditches and flumes were high enough above the mines to furnish a satisfactory pressure for the giants. In rare cases water was pumped from neighboring creeks. The majority of placer mines could be operated from December to May. With the beginning of the winter rains and the later melting of snows, the creeks had increased volume of water, and adequate supply through ditches and flumes could be obtained.

The gravels varied from soft, easily washed material to cemented gravels that had to be loosened by blasting or by jets of high-pressure water. If the gravels had a clayey matrix, the washing was done better by hydraulicking. The quantity of gravels extracted depended on the head used on the giants, but high pressure water was more likely to carry the finer gold out of the sluice boxes.

The Paragon Mine is the only one in the area that was being worked by hydraulicking at the time of this investigation. The work season is from December to May, when adequate water supply is obtained from the Volcano Canyon through ditches and flumes. [4]

Canada Hill

Location. This district is in eastern Placer County 25 miles northeast of Forest Hill and about 10 miles south of Cisco. It includes the Sailor Flat, Robertson Flat, Sailor Canyon and New York Canyon areas.

Geology. The Canada Hill channel, a branch of the Tertiary American River, is believed to have flowed northeast and east across Sailor Canyon and then southeast to join the southwest-flowing main Tertiary channel near French Meadows. The channel is steep, narrow, and has not been too productive. The gravels are usually angular and poorly washed. Bedrock is slate and quartzitic schist of the Blue Canyon Formation (Carboniferous) and schist of the Sailor Canyon Formation (Lower Jurassic). The gravels are capped by rhyolite and andesite. There are some gold-quartz veins that contain abundant sulfides.

Mines. Beauty, Carmac, Lost Emigrant, Merz, Monumental, Pacific Blue Lead, Placer Queen , Reed, Sailor Canyon, Trinidad, Walker, X-Ray.

Colfax

Location. The Colfax district is in southwestern Placer County in the vicinity of the town of that name. It includes the deposits in the Weimar-New England Mills area and placer-mining areas along the American and the Bear Rivers. The Colfax district also has been know as the Illinois district.

History. Placer mining began here soon after the beginning of the gold rush. The locality was first known as Alder Gulch and later Illinoistown. The Rising Sun mine was discovered in 1866. Colfax, which was named for U.S. Senator Schuyler Colfax who visited the place in 1865, was an important center during the construction of the Central Pacific Railroad. Considerable mining continued in the district until about 1900. Small-scale placer mining, some done by skindivers, has continued until the present time.

Geology. Slates of the Mariposa Formation (Upper Jurassic) are found in the central portion of the district, and slate of the Cape Horn Formation (Carboniferous) crops out to the east. A gabbro intrusion lies to the north and northwest, a diabase body to west, and serpentine lenses and amphibolite to the east and southeast. A fairly extensive patch of Tertiary channel gravel is exposed on the north side of Colfax Hill. The extensive quartz gravels in the Bear River are hydraulic mine tailings from the You Bet and Lowell Hill districts.

Ore deposits. The ore bodies occur in a number of northeast-striking quartz veins that usually range from two to five feet in thickness. The ore contains free gold and often abundant sulfides. Considerable highgrade ore was found near the surface during the early days.

Mines. Placer: Bear River Ext., Bear River Tunnel, Burnt Flat, Collins, Rocky Bar.

Damascus

Location. This district is in east-central Placer County at the site of the old town of Damascus, about seven miles southeast of Dutch Flat. It includes the lode mines of the Pioneer-Humbug Bar area on the north and the extensive placer deposits that extend from Damascus south through Forks House to the Sunny South-Gas Hill area.

History. The streams in this area were originally mined during the gold rush, and drift mining began in the late 1850s. The Hidden Treasure drift mine was discovered in 1875 and was worked on a major scale through the early 1900s. There was some mining activity in the district again during the 1930s, and there has been intermittent prospecting since. This area has been quite productive; the drift mines alone have had a total output of more than $12 million and the lode mines several million dollars more.

Geology. An early Tertiary gravel channel extends from Damascus south to Gas Hill and eventually southwest to Michigan Bluff. An intervolcanic channel that apparently eroded away portions of the earlier channel enters the area from Westville to the east. The lower earlier "white" channel is more than 300 feet wide. It contains abundant quartz. The gravels in this early channel yielded 50 cents to $1.75 per ton at the old price, and the gold was coarse. The upper channels are commonly known as "blue" channels. Bedrock is slate and schist, and the gravels are capped by rhyolite and andesite. The channel has been mined by drifting almost continuously from Damascus south to Hidden Treasure, a distance of more than four miles. The quartz veins, which occur in slate, range from two to eight feet in thickness and contain free gold and often abundant sulfides. The ore usually is low to moderate in grade, but the ore shoots had stoping lengths of up to several hundred feet. The Pioneer mine was developed to a depth of 1400 feet.

Mines. Placer: Bullion, Cameron, Comet, Gas Hill, Golden River, Hermit, Hidden Treasure

$4 million, Mountain Chief $700,000+, Mountain Gate, Rainbow, Tickell.

Duncan Peak

Location and History. This is an extensive area of placer deposits in the general vicinity of Duncan Peak and the Greek Store guard station in eastern Placer County. It is 20 miles east of Forest Hill and six miles southeast of Last Chance. The area extends from just south of Duncan Peak south through Duncan Canyon to Ralston Ridge. The area was first mined in the early 1850s. The peak was named for Thomas Duncan, an early-day miner. During these early days, this region supported many Greek placer miners. Intermittent prospecting and development work has continued until the present time.

Geology and Ore Deposits. The placer deposits occur in a complex system of Tertiary channels that extends south and southwest to join the main west-trending Long Canyon channel in the Ralston Divide district. There are a number of tributaries and channel remnants, one of the main ones known as the Chalk Bluff channel. The deposits are up to several hundred feet wide, and were extremely rich in places. Usually the gold is coarse. Quartz is sparse. The bedrock is quartzitic schist and slate, and the gravels are capped by andesite. There are some narrow gold-quartz veins in the district.

Mines. Bald Mountain, Barney, Blue Eyes, Dixie Queen, Glenn Cons., Gold Dollar, Golden Gate No. 1, Hard Climb, Hunted Hole, Miller's Defeat, Pine Nut, Pork and Brown, Red Star, Jack Robinson, Sauer Kraut, Savage, Trap Line, Yellow Jacket.

Dutch Flat

Location. Dutch Flat is in north-central Placer County. This district includes the Alta and Towle areas. The Gold Run district lies just to the south, the You Bet district to the west, and the Lowell Hill district to the northeast.

History. Placer mining began here in 1849. The settlement was established by some Germans or "Dutch" in 1851. Hydraulicking began in 1857 and, during the following few years, the hydraulic mines were highly productive. Operations continued until 1883, when the mines were shut down by anti-debris injunctions. Some work was done in the district again in the 1890s and early 1900s. Logan (1936) estimated the district to have a total output of $4.5 million to $5 million although it may be more. The old town of Dutch Flat is well-preserved and is now a popular tourist attraction.

Geology. This district is located at the junction of several major channels of the Tertiary American River. One channel enters the area from the Lowell Hill district on the northwest, another from Lost Camp and Shady Run on the east, and a third from the Gold Run district on the south. The main channel then continues west and northwest through Little York, You Bet, Red Dog, and Hunt's Hill. It has been estimated that 90 to 105 million yards have been washed here. The gravels have a maximum depth of 300 feet, the lower 150 consisting of coarse blue gravel. The bottom gravels are well-cemented. Bedrock consists of slate, gabbro, quartzite, and amphibolite.

Emigrant Gap

See text under Nevada County, p. 100.

Forest Hill

Location. The Forest Hill district is in south-central Placer County in the general vicinity of the town of that name. This district is fairly large in area and includes not only the "diggings" at Forest Hill but those at Bath to the east, Todd Valley and Dardanelles to the southwest, and Yankee Jim's to the northwest. The district is principally a placer-mining one, although there have been some productive quartz mines.

History. Gold was discovered here in 1850. By 1852 the area was highly productive. In that year the Jenny Lind mine was discovered, and hydraulic mining was introduced at Yankee Jim's by Colonel McClore. The town was an important trading center in those days. By 1868 the mines in the vicinity of the town had yielded more than $10 million. Large-scale hydraulic mining continued until the early 1880s and drift mining until the early 1900s. There was appreciable activity in the district yet again in the 1930s and early 1940s, and a few mines, such as the Paragon and Three Queens, have been worked since. Forest Hill is now an important lumbering center. The total output of the district is estimated to be at least $25 million, and it may be considerably more.

Geology. The main early Tertiary channel of the Middle Fork of the American River enters the district from Michigan Bluff on the east. At Bath it turns north and then west and southwest and continues southwest through Forest Hill. At the Dardanelles mine west of Forest Hill, the channel swings northwest to Yankee Jim's and then north to the Iowa Hill district. An intervolcanic channel extends west-southwest from Baker Ranch to north of Forest Hill. Another intervolcanic channel extends south-southwest between the above and Yankee Jim's. The older quartzitic gravels near bedrock are coarse and well-cemented and have yielded the most gold. Much of the gravel is overlain by rhyolite and-andesite. Bedrock is slate with some phyllite, schist, and serpentine. Some of the gold-quartz veins were rich, especially those that occur near serpentine. The veins are usually three to four feet thick and strike in a northwesterly direction. A number of small but rich pockets were found in the Three Queens mine, the principal lode mine in the district.

Mines. Placer: Baker Divide; Baltimore; Big Spring $150,000; Dardanelles $2 million+; Excelsior; Florida; Georgia Hill, Yankee Jim and Smiths Point, together $5 million; Grey Eagle; Homestake; Independent, New Jersey and Jenny Lind, together $2,653,000; Mayflower $1 million; Maus; Paragon $2.65 million+; Peckham Hill and Todd Valley together $5 million; Pond; San Francisco; Small Hope; Yankee Jim's.

Gold Run

Location and History. This district is in north central Placer County in the vicinity of and south of the town of Gold Run. Extensive Tertiary channel gravels extend from here south to Indiana Hill and the North Fork of the American River. Much of the output in the district has come from the vast Stewart hydraulic mine, which is traversed by U.S. Highway Interstate 80 across its north end. The area was first placer-mined in 1849, and the town was founded in 1854 by O. W. Hollenbeck. The Town was originally called Mountain Springs. From 1865 to 1878 approximately $6,125,000 in gold was shipped from the express office here. Mining on a moderate scale continued until about 1915, with considerable production reported in 1908. There was minor work here in the 1920s and 1930s.

Geology. The deposits are located on a major Tertiary channel of the American River that enters the area from the south and continues north to Dutch Flat. The gravel deposits are more than a mile wide in an east-west direction, three miles long in a north-south direction, and up to 400 feet deep. The lower cemented blue gravel yielded as much as several dollars per yard. The upper gravels contain quartz with clay and sand and averaged 11 to 17 cents per yard, while the top gravels ran to four to five cents per yard. Bedrock is slate in the west portion and gabbroic rock to the east.

Iowa Hill

Location. The Iowa Hill district is in central Placer County in the vicinity of the old mining town of that name. It is an extensive placer-mining district that includes the Roach Hill, Monona Flat, Strawberry Flat, Succor Flat, Grizzly Flat, Shirttail Canyon, and Kings Hill areas.

History. Placer mining began along the American River and its tributaries soon after the beginning of the gold rush. Hydraulic and drift mining apparently began here in 1853, and by 1856 the output was as high as $100,000 per week. By 1880 more than $20 million had been produced from the district. Drift mining continued through the early 1900s, and there was appreciable activity again in the 1930s. Most of the town was destroyed by fire in 1922. The Big Dipper, Occidental and a few other mines have been intermittently worked in recent years. Also snipers and skin divers have been active in the district.

Geology. A main Tertiary channel of the American River crosses the area. There are numerous branches and intervolcanic channels, including the Succor Flat intervolcanic channel, which comes in from the northeast, and the west-trending Morning Star and Grizzly Flat deep channels. The deep channel gravels are well-cemented and in places yielded 1/2 ounce of gold or more per yard. The lowest seven feet were the richest but there also were some rich benches. The bedrock is uneven, and consists of hard slate and phyllite of the Cape Horn Formation (Carboniferous) and amphibolite, which contains a number of deep and rich potholes. To the east the gravels are overlain by thick beds of andesite. There are a few gold-quartz veins in the district.

Mines. Big Dipper $1.2 million, Blue Wing Quartz, Brunn, Buckeye, Campbell, Canyon, Carey, Copper Bottom, Dewey Cons., Drummond, Elizabeth Hill, Excelsior, Fitzpatrick, Gleason $1 million+, Golden Star, Golden Streak, Goodwin, Haymes, H and H, Iowa Hill, Irish and Bryne, Jupiter, Keystone, King's Hill Point, King's Hill Quartz, Lebanon, Mohawk, Morning Star, Occidental, Old Jupiter, Penn Valley, Randal, Roach Hill, United, Strawberry, Twenty One, Welcome, Winchester, Wisconsin Hill, Union.

Last Chance

Location. This extensive placer-mining district is in eastern Placer County in the vicinity of the old mining camp of Last Chance, 10 miles northeast of Michigan Bluff and 15 miles northeast of Forest Hill. It includes the "diggings" here and at Star Town, Deadwood, and American Hill. Last Chance got its name when a starving miner used his last bullet to kill a deer. The mines in-the district were operated almost steadily from the early 1850s until about 1920. There was some activity again in the 1930s, and the El Dorado and Last Chance mines have been intermittently worked in recent years.

Geology. Most of the gold has come from three southwest-trending Tertiary channels. The lowest but youngest is the El Dorado channel, which is steep, only about 100 feet wide, and contains coarse gold. The next youngest is the Sharp Stick channel, which contains clay and coarse boulders. The oldest is the Big Channel which is quartzitic, well-cemented, and up to 800 feet wide. The gravels are capped by andesite. Bedrock is slate and schist of the Blue Canyon Formation (Carboniferous), which encloses some narrow north-striking gold-quartz veins.

Mines. Beaman Ledge, Bear Wallow, Central, Darling, Deep Canyon, Double O, El Dorado,

Elkhorn, Golden Riffle, Grizzly, Harkness, Home Ticket $200,000+, Hornby, Last Chance, Little Hope, Missouri Flat, New Caledonia, Pacific, Pacific Slab, Peters, Rattlesnake, Rublin, Sharp Stick, Star Town.

Lincoln

Lincoln is in western Placer County, 15 miles west of Auburn and 11 miles north of Roseville. The town was named for Charles Lincoln Wilson, who built the California Railroad here in 1861. During the 1930s considerable amounts of gold were recovered by dragline dredging from gravels in lower Auburn Ravine, just east of Lincoln, and in Doty Ravine, a few miles to the north. This was probably the most profitable dragline dredge field in the state. Recoveries ranged from 15 to as high as 60 cents per yard. Digging depths ranged from 5 to 20 feet. The gravels are underlain by soft tuff. The total dredged area is about 1200 acres.

Michigan Bluff

Location. The Michigan Bluff district is in south-central Placer County. It is best known as a placer mining district, and includes the Turkey Hill, Byrd's Valley, and Baker Ranch areas. The Damascus district is to the north and the Forest Hill district is to the west.

History. The town, first settled in 1850, was originally known as Michigan City. In 1858 the land began to slide into the river, so the town was moved higher up on the mountain side and became Michigan Bluff. Hydraulic mining began here in 1853, and the district soon became highly productive. During the middle and late 1850s, the gold output averaged $100,000 per month. Leland Stanford, Governor of California and one of the builders of the Central Pacific Railroad, operated a store here from 1853 to 1855. His old home still stands. Activity in the area declined during the 1870s, but some work continued intermittently through the early 1900s and again in the 1930s. Much of the region was devastated by fire in 1960.

Geology. This district is at the junction of two major Tertiary channels, one that comes in from the north from the Damascus district and the other comes in from the southeast from Ralston Divide. Just to the north at Baker Ranch there is an intervolcanic channel. The lower gravels at Michigan Bluff are nearly pure quartz with many large boulders. The gravels were extremely rich, the gold yield from six million yards reportedly having been $5 million. Much of the gold was coarse. Bedrock is slate and schist, and to the west there is serpentine. Some narrow gold-quartz veins are present.

Mines. Placer: Adams, Anna Sue, Baker Ranch, Beehive, Big Gun (Michigan Bluff) $1 million+, Bowen, Bower, Britt, Buckeye, Burnham, Burns, Burroughs, Drummond, Eastman, El Dorado Hill, Franklin, Golden Chief, Golden Gate, Gorman, Hazard, Hoffman, Imperial, Lightfoot, Manhattan, Mary Anna, Rainbow Land, Russel, Sage Hill, Turkey Hill, Washburn, Washington, Weeks, Weske.

Ophir

Location. This district is in southwestern Placer County in the vicinity of the old town of Ophir. It extends east to Auburn and west to Gold Hill and includes the area known as the Duncan Hill mining district.

History. Gold-bearing surface gravels were discovered at Ophir and Auburn in 1848 and for several years yielded substantial amounts of the mineral. The quartz veins were then developed, and the Ophir and Duncan Hill districts were organized. Appreciable amounts of high-grade gold ore were recovered during the 1860s, 1870s, and 1880s, but mining activity in the district declined after that. The mines were active again from the early 1900s through the 1930s with substantial production, but little has been done since. The value of the total output of the district is estimated at more than $5 million.

Geology. The mineralized zone is on the northeast flank of a granodiorite and quartz-diorite stock that is intrusive into amphibolite schist. A series of west-northwest-striking and south-dipping quartz veins occur in the granitic rocks or along the granitic rock-amphibolite schist contact. A few veins are in the amphibolite. The ore contains free gold in places with often abundant pyrite, galena, and chalcopyrite; much of the ore is base. Milling ore ranged from 1/4 to one ounce of gold per ton. Considerable high-grade ore was taken close to the surface during the early days. The veins range from one to five feet in thickness, and several were mined to depths of more than 1000 feet. The ore shoots had stoping lengths of up to 250 feet.

Mines. Belmont, Black Ledge, Centennial 150,000+, Conrad $50,000, Crater $750,000, Doig, Eclipse $100,000+, Gold Blossom $216,000+, Grass Ravine, Green $150,000+, Green Emigrant $ 150,000+, Hathaway $336,000, Julian, Mina Rica $55,000, Moore $180,000, Oro Fino $500,000+, Pine Tree, Rock Creek $200,000?, St. Lawrence, St. Patrick $149,000+, Three Stars $415,000.

Penryn

Location. This district in southwestern Placer County in the vicinity of the town of Penryn, also has been known as Stewart's Flat. The Ophir district adjoins it on the north and northeast. The town of Penryn was named by Griffith Griffiths, who opened granite quarries here in 1860, after Penryn, Wales. Appreciable mining activity here during the 1930s and early 1940s was highlighted by work at the Alabama, Chicago, and Sicily mines.

Geology. The principal country rocks are granodiorite and quartz diorite with some diorite and gabbro. A number of north- to northwest-striking quartz veins, cropping out with nearly vertical dips, range from one to five feet in thickness. The ore contains free gold with varying amounts of sulfides. Argentite and tellurides are occasionally present. The milling grade ore usually averaged 1/3 to 1/2 ounce of gold per ton, but some was considerably richer.

Mines. Alabama $1 million+, Chicago $100,000+, Elizabeth, Highway 40, Jenny Lind, Mary Len, Penryn, Sicily $100,000+.

Ralston Divide

Location. This is a placer-gold district in southeastern Placer County 25 miles east of Forest Hill. It is an extensive region that includes the Ralston Ridge, Long Canyon, and Nevada Point Ridge areas. It is just south of the Duncan Peak district.

Geology. The deposits are along a west- and southwest-trending Tertiary gravel channel known as the Long Canyon channel, the eastward continuation of the main Michigan Bluff-Forest Hill channel. The gravels are interbedded with rhyolite tuffs and contain granitic boulders. The main Long Canyon channel is fairly broad and flat and covers large areas, but it is generally of low grade. The gold usually is fine. Bedrock is slate and quartz-bearing schist of the Blue Canyon Formation (Carboniferous), with granodiorite to the east. The gravels are capped by andesite and rhyolite. Mining was done by hydraulicking and drifting.

Mines. Blacksmith Flat, Clydesdale, Goggins, Granite, Ralston, Russian Ravine, Zuver.

Rattlesnake Bar

See text under Eldorado County, p. 59.

Placer County Dredging Operations

• There is only one dredging company operating in Placer County. The property of this company is located in Placer and Yuba counties, near the town of Wheatland, along the Bear River, which drains an area of about 287 square miles between the Yuba and American Rivers. The headwaters of the Bear River do not reach back to the crest of the range, so that it seldom receives precipitation under form of lasting snow. It is torrential in character, having no forested areas except in its upper portion. At the head of the Bear River ancient gravel deposits are found.

Bear River Mining Company began operations in 1901, and at the present time has an operating plant consisting of one dredge. The holdings of the company comprise an area of 933.10 acres, located in the Bear River Basin, 5 miles from the town of Wheatland, in sections 37 and 30, township 14 north, range 6 east, in Placer County, and in sections 27 and 34, township 14 north, range 5 east, in Yuba County.

Some of this land had been mined previous to dredging, and none of it as ever under. The property was originally owned by R. D. Evans, of Boston, and W. P. Hammon, of San Francisco, who organized the Bear River Exploration Company, and later the Bear River Mining Company, which took over the holdings of the first company. [2]

Rocklin

Location. This is a placer-mining district in southwestern Placer County, two miles east of Rocklin and two miles south of Loomis. A gravel channel of the Tertiary American River trends southwestward through the area. There are actually two channels: an upper loosely consolidated intervolcanic channel that contains some gold and a lower well-cemented quartz-rich channel that in places was rich. The lower gravels yielded $1 or more per yard in gold. The gold is fine, flat, and sometimes rusty. The lower channel is as much as 1500 feet wide. Bedrock is granodiorite, and in places the gravels are capped by andesite. The Lee drift mine, one of the principal sources of gold in the district, has been prospected in recent years.

Volcanoville

See text under El Dorado County, p. 59.

Westville

Location. This placer-mining district is in eastern Placer County about 17 miles northeast of Forest Hill and 10 miles due south of Emigrant Gap. It adjoins the Damascus district on the west and the Canada Hill district on the east. The district includes the Macedon Ridge, Whiskey Hill, and Secret Canyon areas.

Geology. The placer deposits lie along a Tertiary intervolcanic channel of the American River, known as the Red Point channel, which extends southwest into The Damascus district. A tributary known as the Whiskey Hill or Black Canyon channel joins the Red Point channel at Westville. This tributary is narrow, steep, and contains coarse gold. The gravels are capped by andesite, and bedrock is quartz-bearing schist and state. Much of the development in this district has been drift mining.

Mines. Golden Fleece, Greek, Green, Hogsback, Herman, Hungry Hollow, Macedon, Osborne, Union.

Wheatland

This is a small placer-mining district in the vicinity of the town of Wheatland on the lower Bear River in western Placer and southern Yuba Counties. During the gold rush placer gold was recovered from the creeks and streams. During the 1930s gold was recovered from the gossan by cyanidation at the Dairy Farm copper mine a few miles to the east and by dragline dredging in some of the ravines.

• Placer County Gold Placer Mines–Foresthill Quadrangle

Adams Pit. Sec. N1/2-27, T. 14N, R. 10E. On divide between First and Second Brushy Canyons. North part worked by hydraulicking.

Big Dipper. Sec. 3 & 4, T. 14N, R. 10E. Worked by hydraulicking 1858-1882, then by drifting until connection made with Calif. Morning Star Mine. Gravel said to have averaged $6.00 per ton.

Big Spring. Sec. 33 & 34, T. 14N, R. 10E; Sec. 3, T. 13N, R. 10E. On west side of Dardanelles mine. Opened in 1854 as drift mine and worked until late 1800s. Part of work done through 1400' Big Spring adit, and part through adit shared with Dardanelles. In 1882 reported to have had production of $150,000. Pay gravel 5' thick, well cemented. Much of property now subdivided into town lots.

Buckeye. Sec. 11 & 14, T. 14N, R. 10E. Drift mine. Adit southeast into hillside. Mining confined to boulder bed resting on bedrock.

Campbell. Sec. SE1/4-4, T. 14N, R. 10E. Two adits run N.80°W. and N.40°W. in bedrock. Remains of milling plant, concentrator, sluice boxes.

California Morning Star. Sec. 33 & 34, T. 15N, R. 10E. Drift mine. Portal and mill east side of Indian Creek. Drift driven eastward into bedrock below channel, gravels worked through raises. Connected with drift from Big Dipper. Average width of channel 200-300'. Pay gravel extended several inches into slaty bedrock. Part of placer deposit worked by hydraulicking. Average yield said to be $7.00/ton.

Copper Bottom. Sec. 25, T. 15N, R. 10E. Upper run or bench deposit about 50' above main channel worked in 1922. Deeper channel reached through an adit 400' and a slope 300', requiring a lift of 120' for water.

Dardanelles (Oro & Dardanelles). Sec. 34 & 35, T. 14N, R. 10E; Sec. 2, T. 13N, R. 10E. One of major gold sources in Foresthill area. Worked almost continuously by drifting and hydraulicking from 1853 to 1890. By 1890 credited with total output of $2,000,000. Intermittently worked by drifting from 1890 to 1930. Main source of production was well-cemented, 5' lower gravel overlain by coarse gray sand. Mined to average width of 75'. Property now site of sawmill and landing strip of Stockton Box Co.

Elizabeth Hill. Sec. SE1/4-5, T. 14N, R. 10E. Drift mine NE part of King's Hill, later worked by hydraulicking. Tunnels and shafts probably used to supply water from nearby creek or wells. Said to be exceptionally rich in coarse gold.

Gleeson Diggings. Sec. SE1/2-4, T. 14N, R. 10E; N1/2-9, T. 14N, R. 10E. Placer deposit, extensively hydraulicked. Several adits at eastern margin: N.75°W.; N.55°W.; S.E.; all caved. Water supply for hydraulicking presumably from Refuge Canyon through tunnels in bedrock.

Golden Streak. Sec. 35, T. 15N, R. 10E. Drift mine. Shaft driven in mudflow capping gravel. Closed in 1951.

Goodwin. Sec. S1/2-3, T. 14N, R. 10E. Drift mine on western flank of Prospect Hill. Two adits, one N.75°W. Both caved. Diggings south and northwest of portals. Large hydraulic pit.

Trenches at regular intervals indicate systematic sampling.

Grey Eagle. Sec. 6, T. 13N, R. 10E. Portal of main adit runs south in bedrock on south side of Owl Creek. Channel near Spring Garden extensively prospected 1915-16. Considerable amounts of cemented gravel run through old stamp mill, but average returns evidently not satisfactory.

Haymes. Sec. 8, T. 14N, R. 10E. Drift mine. Two adits driven, the older trending north; other N.40°W. along Ione gravels resting on bedrock. Problem to keep roof from caving.

Hazard & Weske. Sec. NE1/2-20, T. 14N, R. 11E. Worked for about 3000', from Weske mine. Incline was run downstream several hundred feet into Muir Tunnel claim.

H &H (Hendrickson & Hoover). Sec. 18, T. 14N, R. 10E; S1/2-7, T. 14N, R. 10E. Drift mine on NW flank of King's Hill; closed in 1948. Two adits and some open pits. Eastern adit runs N.15°W. about 720'. Filled with water in 1951. Other adit blocked with boulders. In east drift, gravel bed rests on bedrock, underlies mudflow, is 1-1/2-2' thick, consists of slate and metavolcanic rock. Rounded white quartz gravels of Ione type rare.

Independent, New Jersey, and Jenny Lind. Sec. W1/2- 36, T. 14N, R. 10E; E1/2-35, T. 14N, R. 10E. Claims under town of Foresthill. Primary stream deposit about one mile in length, flowing west, between canyon slope and deep blue lead. Expectation of finding western segment of channel, beyond point where it presumably had been cut off by deep blue lead inspired work after pioneer miners.

Iowa Hill. Sec. 27,33,34, T. 15N, R. 10E. Originally a drift mine, worked by hydraulicking before 1880, along Indian Canyon front. Main adit runs SE. Worked intermittently until 1935 by Iowa Hill Gold Mining Co. Reopened in 1940 and 1944. Probably yielded more than $10,000,000 between 1849 and 1911.

Jupiter. Sec. 2, T. 14N, R. 10E. Drift mine, portal SE of Grizzly Flat. First part of drift trends NE in bedrock, then follows trough of channel (about 4500'). Crosscuts on both sides. Gravel bed 1-1-1/2' thick contains rounded/subrounded pebbles of quartz and chert—underlies mudflow and overlies bedrock. Some places andesite mudflow rests on bedrock, other places bedrock boulders underlie mudflow. Worked 1937-42, by a Mr. Carlson. Except for a few rich pockets, gold reported sparsely distributed in pebble bed overlying bedrock.

Keystone. Sec. N1/2- 22T. 14N, R. 10E. Sec. S1/2-15, T. 14N, R. 10E. Hydraulic pit. Rhyolitic sands and gravels overlie basal boulder bed of probable Eocene age.

Little Indian Creek. Sec. 26,27, T. 15N, R. 10E. Drift mine. Western adit runs N.60°E. into SW side of Strawberry Flat. Drift driven along bed of rhyolitic sand and gravel overlying bedrock and underlying mudflow. (Intervolcanic channel?) Upstream adit driven from creek bed N.60°W. into bedrock. Water-filled.

Mayflower. Sec. 23,24, T. 14N, R. 10E. Mine first worked by hydraulicking, then drifting. Same lead as bottom lead in Paragon mine. Gravel 2-14' thick yielded average of $7.00 ton. Cemented and needed milling. Upper lead, 150' above main channel said to have produced much gold. In hydraulic pit no adit can be seen, no bedrock exposed.

Michigan Bluff. Sec. 21, T. 14N, R. 11E. In western ends of mines auriferous gravels are about 40' thick, reported profitable. Total yield 1849-1911 about $5,000,000.

Occidental. Sec. 3, T. 14N, R. 10E.; Sec. 34, T. 15N, R. 10E. Drift mine owned by a Mr. Carlson. Inclined shaft sunk NNW, 2 levels 60' apart. Lower level follows trough of channel with boulders of bedrock. Coarse gold reported in pockets. Upper level driven in thick gravel bed of rounded/subrounded pebbles, cobbles, boulders-mostly quartz-intercalated with rhyolitic sands containing abundant biotite. Extensive gravel beds on both

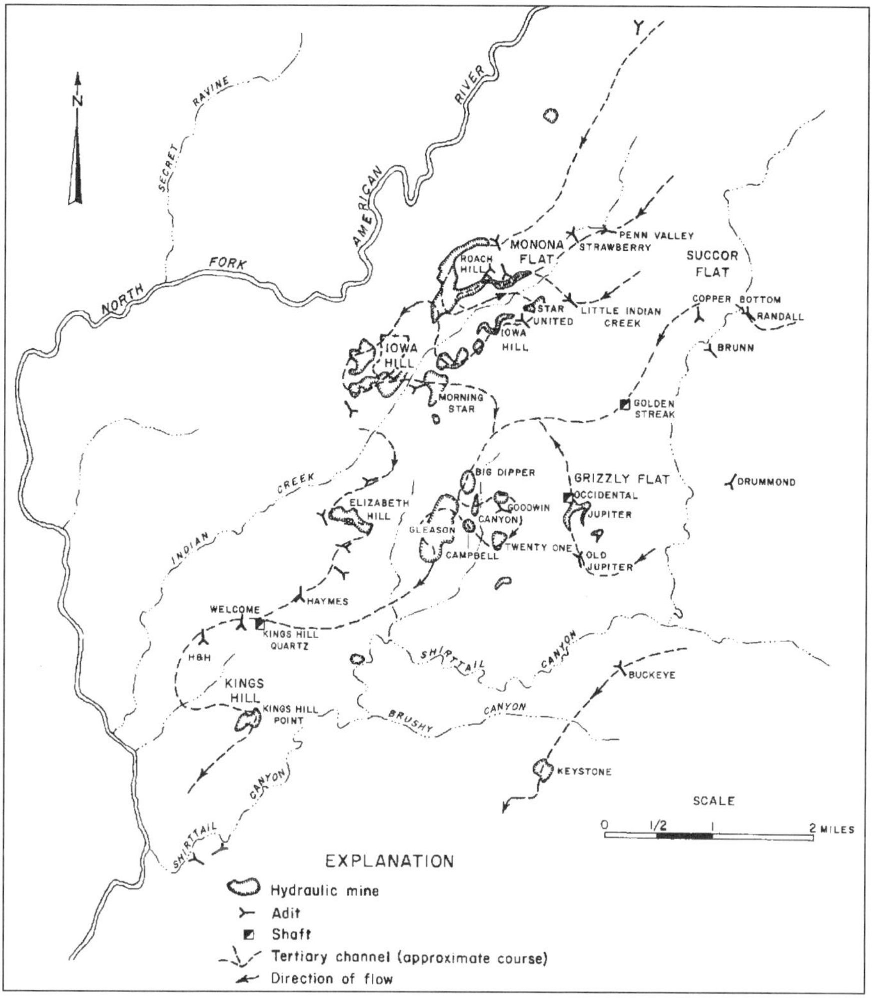

Map of Iowa Hill District, Placer County, showing locations and Tertiary channel courses.

sides of the level appear to be flood-plain deposit. Paystreak in ferruginous conglomerate associated with ferruginous shale parting. Gold recovered from upper level reported finer than that from lower. Mine has small milling plant.

Old Jupiter. Sec. 11, T. 14N, R. 10E. Drift mine on west flank Long Point Hill. Adit running S.30°E. has caved at portal.

Paragon. Sec. 19, T. 14N, R. 11E; Sec.30, T. 14N, R. 11E; Sec. 24, T. 14 N, R. 10E. Hydraulic mine. Produced $2,651,000 prior to 1902. Pay reported to be mainly in lower 150' (between upper Paragon lead and bedrock). Lower and upper leads worked by drifting and then by hydraulicking (Mr. William Wilson) during winter when adequate water available from Volcano Canyon. Said to have been profitably worked recently. Large amount of fine gold reported lost with heavy minerals in sluice boxes.

Penn Valley. N1/4-26, T. 15N, R. 10E. Drift mine on south side of Indian Creek. Adit runs through hard metavolcanic rock so timbering unnecessary. 1500' drift runs SE. Two short raises to metavolcanic and serpentine gravel and boulder beds. Quartz gravels nearly absent. Mine intermittently worked by a Mr. Henderson.

21 Placer. Sec. S1/2-3, N1/2-10, T.14N, R. 10E. Old drifts not worked for 50 years. Part of placer deposit has been mined by open-cut method.

Pond Pit. Sec. 3&4, T. 13N, R. 10E. In Todds Valley. Extensively worked before 1875 and proved to be a large producer. Deposit could not be worked further because of increasing overburden. Drifting not feasible because of low gold content.

Randall. Sec. 25, T. 15N, R. 10E. Drift mine. Old adit 600' long reported to run SE into hill; new one N.30°W. Mining confined to thin rhyolitic gravel bed overlying bedrock and underlying mudflows; apparently on a small intervolcanic channel.

Roach Hill. Sec. 27, 28, T. 15N, R. 10E. Drift and hydraulic mine. Bedrock exposed between tailings. One drift on west flank of hill worked intermittently on small scale.

San Francisco. Sec. 33 & 34, T. 14N, R. 10E. Drift and hydraulic mine between Pond mine on west and Big Spring mine on east. Chief source of gold was well-cemented quartzitic gravel in slate bedrock.

Small Hope. Sec. 27, T. 14N, R. 10E. Drift and hydraulic mine on north side of First Brushy Canyon. Adits driven to prospect deposit of 30-40' gravels, sands, shale overlain by andesite mudflow. Channel also contained cemented gravels of intervolcanic period.

Star United. Sec. 27, T. 15N, R. 10E. Old mine fronting on Indian Canyon 1-1/2 mi. E of Iowa Hill Mine. Extensively mined by drifting and hydraulicking.

Strawberry. Sec. 26, 27, T. 15N, R. 10E. Drift mine under lease to M. Capurro in 1950. Portal on S. bank of Indian Creek. Drift runs SE about 1200'. Follows trough of small channel. Mining confined to nonpersistent rhyolitic gravel bed overlying bedrock and underlying mudflow. Channel evidently intervolcanic.

First Sugar Loaf. Sec. 32,33, T. 15N, R. 10E. Eastern half extensively, worked by drifting and hydraulicking. Western half not hydraulicked owing to thick overburden of mudflow andesite.

Volcano. Sec. 19, T. 14N, R. 11E. Adit runs north in gravel bed overlying serpentine bedrock and underlying mudflow.

Welcome. Sec. 7, T. 14N, R. 10E. Drift mine on south flank of King's Hill. Westernmost adit trends N.15°W. Eastern adit trends N.20°W. for 700' but is caved. New adit being driven parallel to latter in harder bedrock in August 1951. Mine equipped with big water pump, small mill, and concentrator. [4]

• Below are a few notes on recent operations and on a few properties that are believed to contain reserves of possible commercial grade.

Duffy property, in sec. 3, T. 13 N., R. 11 E., and sec. 24, T. 13 N., R. 9 E., and intermediate sections. The gravel has been sampled for a distance of 1-1/2 miles from a point near the mouth of Volcano Canyon near the line between sec. 5 and 6, T. 13 N., R. 11 E., and going down stream. The depth ranged from 20 to 25 feet. Width of the gravel is 350 to 400 feet. Eight miles at the lower end of the holding were sampled by means of sinking caissons by hand. Many of these did not reach bedrock, as it was possible to sink them to a depth of only 15 to 16 feet. Gravel is 500 to 700 feet wide on this lower end of the holding. A part of Duffy's holdings known as Horseshoe Bar, in secs. 4, 5, T. 13 N., R. 11 E., was worked with a dragline dredge in 1941 and 1942.

Lebanon Consolidated Mines, San Francisco, worked the Occidental drift mine in the Iowa Hill district from January 1 to December 31, 1941. The yield from 3766 cubic yards of gravel was 536 ounces of gold and 63 ounces of silver.

Lost Camp mine is a hydraulic mine of 440 acres 2 miles by road from Blue Canyon, at an elevation of 4300 feet, in secs. 22, 23, T. 16 N., R. 11 E. The channel contains interbedded layers of soft rhyolite tuff and free-washing white quartz gravel. Work on the property has proceeded at intervals over a long period by three different methods, ground sluicing, hydraulicking, and drifting. Several million cubic yards of gravel probably remain unworked, but little is known about the gold content.

Mayflower Gravel Mining Company, San Francisco, holds a very large mining property at Foresthill now containing 5800 acres in secs. 15, 22, 23, 24, 25, 26 T. 14 N., R. 10 E., and including the following claims: Texas, Sacramento, Washington, Garland Mill Slope, Excelsior, Hope, Uncle Sam, Green Spring, Live Oak, Small Hope, New Jersey, Brushy Slide, Dardanelles, Oro, and Adams pit. The Mayflower ground contains an unworked section of Big Blue Lead 1-1/2 miles long including those parts in the Excelsior and Garland Mill Slope claims.

McGeachin Placer Gold Mining Company has extensive holdings of placer gravel in secs. 2, 3, 4, 10, T. 14 N., R. 10 E., near Iowa Hill. These properties include 1700 acres. Ground suitable for hydraulic mining is known as the Bog Dipper. Other claims include the Jupiter, Hazelroth, Schwab, Weber, and Winchester. A report was made in December 1938 by F. H. Reynolds and I. E. Rose. Sampling was done in old hydraulic banks roughly 50 feet in height and consisted of cuts 2 feet wide and 1 foot deep. Rose thinks that the ground contains 25,000,000 to 30,000,000 cubic yards of gravel without overburden, in a roughly circular area 3500 feet long and more than 2000 feet wide. Additional yardage is available with overburden.

Paragon mine in secs. 13, 18, 19, 30, 31, T. 14 N., R. 11 E., and sec. 24, T. 14 N., R. 10 E., is one of the large hydraulic mines of the Forest Hill district 2 miles from Foresthill at a place formerly called Bath. The property now contains 1760 acres.

Stewart Gravel Mines, Auburn, controls about 3800 acres on the great Eocene river channel that passes through Dutch Flat and Gold Run. The property is in secs. 2, 3, 4, 9, 10, T. 15 N., R. 10 E., and covers a length along the channel of 2 miles. It is conspicuous because Highway 40, the main route from California to Truckee, runs for a mile at the base of one of the old hydraulic banks nearly 200 feet high. The main line of the Southern Pacific Railroad is at the top of this old hydraulic face. The channel is 1-1/2 miles wide at this point and the gravel was originally 500 feet deep in the middle of the channel, so a depth of several hundred feet remains unworked [1]

PLUMAS COUNTY

Butt Valley

Location. This district is in northwestern Plumas County. It is an extensive area that lies between Lake Almanor on the north and the Virgilia-Twain area on the south and southwest. The district includes the Caribou and Seneca areas. The Crescent Mills district is just to the east, and the Meadow Valley district lies to the south.

History. During the gold rush vast amounts of placer gold were recovered from the Feather River and its tributaries. The valley was named for Horace Butts, a successful early-day miner. The town of Seneca was an important center at that time. The Butt Valley Reservoir was constructed by the Great Western Power Company in 1921. Lode-gold mining, which originally began in the 1850s, continued through the 1930s. Since then there has been intermittent prospecting at a few properties such as the Sunnyside mine. Skin divers and weekend prospectors have been active along the streams.

Geology. The district is underlain by a series of metamorphosed sediments ranging from Silurian to Triassic in age. Slate is most abundant, but also present are sandstone, limestone, quartzite, and conglomerates. In addition, there are greenstones and serpentine.

Ore deposits. The gold-quartz veins occur in brecciated zones in all of the formations, but they are most abundant in the slates and greenstones. The ore contains free gold and varying amounts of sulfides, including arsenopyrite. Milling ore averaged usually only a few dollars per ton, but the ore bodies were extensive. Some of the veins and vein zones are 20 or more feet thick. An appreciable number of high-grade pockets were found. In places the placer deposits, both Tertiary and Recent in age, were extremely rich.

Mines. Placer: Barker Hill, Cameron $100,000+, Dominion, Ellis, Lot, Providence Hill, Red Rock, Sunnyside $150,000+, Swiss, Dutch Hill $575,000+.

Crescent Mills

Location. This district is in north-central Plumas County. It contains a northwest-trending belt of lode and placer deposits that extends from the vicinity of the town of Crescent Mills northwest through Greenville to Almanor, a distance of about 10 miles. At one time it was also known as the Cherokee district. Crescent Mills has been the most productive of the districts in the northern end of the Sierra Nevada.

Geology. This area is underlain by a series of northwest-trending beds of Paleozoic and Mesozoic metamorphic rocks. They are: pre-Devonian metarhyolite; Silurian metasedimentary rocks; Taylorsville Formation (Devonian) slate and sandstone; Taylor Meta-andesite (Carboniferous); Calaveras Formation (Carboniferous to Permian) metasedimentary rocks (four formations); greenstones (Carboniferous); Triassic metasedimentary rocks; Jurassic granodiorite and serpentinized basic and ultra-basic intrusives. Overlying the bedrock are patches of Tertiary and Quaternary gravels.

Ore Deposits. A number of northwest-striking quartz veins occur principally in the greenstones, slates, and granodiorite. Some of the veins are as much as 20 feet thick. The ore contains free gold, auriferous pyrite, and smaller amounts of other sulfides. The surface ores were especially rich, and numerous highgrade pockets were encountered during the early days. Milling-grade ore usually contained slightly less than 1/4 ounce per ton, but the ore

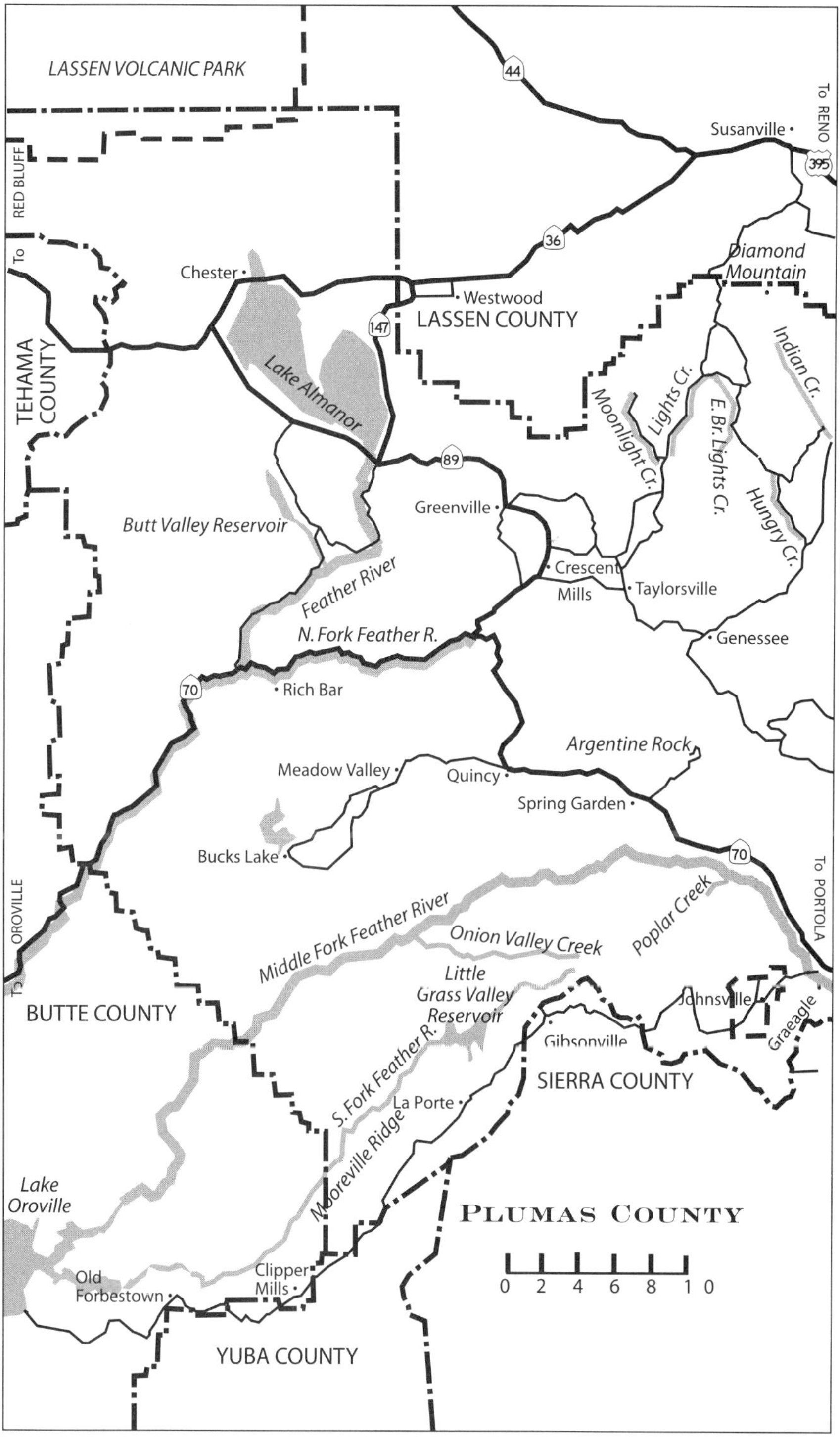
LASSEN VOLCANIC PARK
To RED BLUFF
TEHAMA COUNTY
Chester
Westwood
LASSEN COUNTY
Susanville
To RENO
Diamond Mountain
Lake Almanor
Moonlight Cr.
Lights Cr.
E. Br. Lights Cr.
Indian Cr.
Hungry Cr.
Greenville
Butt Valley Reservoir
Feather River
Crescent Mills
Taylorsville
Genessee
N. Fork Feather R.
Rich Bar
Argentine Rock
Meadow Valley
Quincy
Spring Garden
Bucks Lake
To OROVILLE
To PORTOLA
Middle Fork Feather River
Onion Valley Creek
Poplar Creek
Little Grass Valley Reservoir
BUTTE COUNTY
Johnsville
Graeagle
Gibsonville
SIERRA COUNTY
S. Fork Feather R.
La Porte
Moreville Ridge
Lake Oroville
PLUMAS COUNTY
0 2 4 6 8 10
Old Forbestown
Clipper Mills
YUBA COUNTY

shoots had stoping lengths of up to 300 feet. Iron and copper also occur in the district.

Mines. Altona, Arcadia, Cherokee $250,000, Crescent $500,000+, Dagian, Droege $300,000+, Gold Stripes, Green Mountain $1 million to $2 million, Hobson, Indian Falls, Indian Valley $1.8 million+, Leete, Long Valley, Monitor, Mountain Lily, New York $400,000+, Pearless, South Eureka, Wardlow, Whitney.

Genesee

Location. The Genesee district is in the southeast end of the Crescent Mills-Taylorsville-Genesee gold belt of east-central Plumas County. This is not a single belt but rather several parallel belts or zones of gold and copper mineralization. The well-known Walker copper mine is in this district.

History. Gold was first placer-mined in the streams during the gold rush, and lode mining began soon afterward. Genesee is believed to have been named by the Ingalls family for a valley in New York State. Mining activity continued almost steadily through the early 1900s. The Walker copper mine was worked on a major scale from about 1915 to 1942, and the concentrates were delivered to the Western Pacific Railroad at Spring Garden via a nine-mile aerial tramway. There has been intermittent gold and copper prospecting in the district since.

Geology. This area is underlain by the same series of Paleozoic and Mesozoic metamorphic rocks found in the Taylorsville district to the west and northwest (see Taylorsville district). Contact metamorphism, especially in the vicinity of the Walker mine, has altered the rocks into hornfels and schist. The gold-ore deposits consist of either quartz veins or zones of quartz stringers that contain free gold, limonite, and sulfides. A number of high-grade pockets have been found. There are several patches of auriferous Tertiary gravels.

Mines. Austrian Syndicate, Big Cliff, Blue Bell, Bullion, Calman, Cosmopolitan, Five Bears, Green Ledge, Gruss $460,000, Hinchman, Magpie, Mountain Lion, Native Son, Peter, Taylor (placer), Wards (placer).

Gibsonville

Location.. This placer-mining district is in northwestern Sierra County and southern Plumas County in the vicinity of the old town of Gibsonville. It is about six miles northeast of La Porte and 20 miles due north of Downieville. The district includes the Whiskey Diggings, Hepsidam, and Bunker Hill areas. The area was first worked during the gold rush, and considerable drift mining and some hydraulic mining followed in 1875-95. There has been intermittent prospecting since.

Geology. The deposits are located on the Tertiary North Fork of the Yuba River, which is also known as the La Porte channel. The channel extends in a southwest direction through the district. The channel gravels are as much as 1500 feet wide and capped by thick beds of andesite and clay. The gold was found mostly in quartz-rich gravels near bedrock and was mostly coarse-grained. Some ground yielded up to $3 per yard of gold at the old price. There are a number of drops in the channel, caused possibly by bedrock faulting. The bedrock consists of amphibolite with serpentine, slate, and schist lying to the east. The channel has been drift-mined almost continuously from Hepsidam southwest through Gibsonville to the Thistle shaft, a distance of about five miles. Although the district is reported to have had only a moderate output, extensive drifting and a number of rich pay streaks indicate that it must have been much more productive than originally believed.

Mines. Bellevue group, Bunker Hill, Empire, Feather Fork (Thistle), Garnet, Gibsonville

Water and Mining (hydraulic), Homestake, Taber (Hal Taber), Union-Keystone $600,000, Washington (North American Cons.) (hydraulic), Whiskey (hydraulic).

Granite Basin

See text under Butte County, p. 16.

Johnsville

Location. The Johnsville district is in south-central Plumas County. Both a lode- and placer-gold district, it is at the north end of a major belt of gold mineralization that extends southward through the Sierra City district in Sierra County.

History. The river and stream gravels in the general area were first placer-mined in 1849 or 1850. The Eureka quartz vein, discovered in 1851, quickly brought many miners to the region, and most of the area was soon covered with claims. Considerable coarse gold was recovered from the creeks and considerable high-grade ore from the lode mines during those early years. Both the Plumas-Eureka and the Jamison mines were operated on a major scale until the early 1900s, ,when mining activity in the area declined. Johnsville was named for William Johns, manager of the Plumas-Eureka mine. There was intermittent activity in the district from the period of World War I until around 1943.

The area suffered from a number of disastrous fires. Part of the town and some of the mines became Plumas-Eureka State Park in 1959. The value of the total output of the district is unknown, and there have been a number of extravagant claims. The author estimates the production to be somewhere between $10 million and $20 million. This was a well-known early-day "snowshoe" or ski resort area.

Geology. A considerable variety of rocks crops out in this district, including north- and northwest-trending belts of slate, schist, quartzite, and limestone on the west; metadacite or quartz porphyry to the south; a gabbroic intrusion in the central portion; and greenstone to the east. Portions of the region are overlain by Tertiary andesite. Much of the central portion of the area is covered with glacial detritus. A number of patches of Tertiary gravels yielded gold in the early days. Massive bodies of magnetite are found to the west.

Ore Deposits. There are a number of north- and northwest-trending quartz veins and several wide complex systems of quartz veins. The individual veins usually are only a few feet thick. These contain free gold and often abundant pyrite and varying amounts of galena, chalcopyrite, and arsenopyrite. A number of high-grade pockets were taken from near the surface in the early days. Milling-grade ore contained from a few dollars to more than one ounce gold per ton. The sulfide concentrates sometimes held more than $150 in gold per ton. The ore shoots had horizontal stoping lengths of as much as several hundred feet.

Mines. Placer: Beckwith Cons. drift, Continental drift, Queen drift, Brown Bear hydraulic.

La Porte

Location. This district is in southwestern Plumas County in the general vicinity of the old mining town of La Porte, 25 miles south of Quincy and 56 miles northeast of Oroville. It was one of the great placer mining districts of the state.

History. The streams were placer-mined early in the gold rush and were reported to have had very rich yields. The town, first known as Rabbit Creek, was renamed in 1857 after La Porte, Indiana. Hydraulic mining began in the middle 1850s and continued through the 1880s. During this time the district was enormously productive; the output from 1855 to 1871 alone was reported to have been at least $60 million. Appreciable drift mining and some lode mining were carried on. Some mining activity continued until the period of World War I. The district was prospected again during the 1930s, but apparently little mining has been done here since. La Porte was a noted early-day "snowshoe" or ski resort.

Geology. The main Tertiary channel of the North Fork of the Yuba River, known as the La Porte channel, extended south-southwest from Gibsonville into this district. From here the channel continued southwest and south again to be joined by a branch from the east from the St. Louis-Table Rock area. The main channel today continues on south to the Poverty Hill and Brandy City districts. At La Porte, the channel is 500 to 1500 feet wide and as much as 500 feet thick. The lower gravels are quartz-rich and up to 80 feet thick. Most of the gold was recovered from near bedrock. The gravels are capped by thick beds of sand and "pipe" clay. During the heyday of mining in the district, these lower gravels yielded from 1/10 to as much as one ounce of gold per cubic yard. To the east the channel deposits are capped by andesite as much as 800 feet thick. Also to the east, considerable faulting has disturbed the channel gravels. Bedrock consists principally of amphibolite, with a one-mile wide belt of slate and quartzite of the Calaveras Formation (Carboniferous to Permian) in the central portion. There are some narrow gold-quartz veins in the district.

Light's Canyon

Location and History. This district is in northeastern Plumas County in the general area of Light's Canyon. It was named for Ephriam Light, a pioneer rancher. It includes the Moonlight Valley, Indian Valley, Engelmine, and Kettle Rock areas. It is also an important copper-mining district; the Engels and Superior mines are located here. Placer mining was originally done in the district during the 1850s and continued through the early 1900s. The Lucky S lode mine has been prospected in recent years. For some years the district was served by the Indian Valley Railroad.

Geology and Ore Deposits. The central portion of the district is underlain by granodiorite and quartz diorite. To the south are various metamorphic rocks of Paleozoic and Mesozoic age similar to those found in the Taylorsville district (see Taylorsville district). North and east are Tertiary gravels and volcanic rocks. Both the Recent and Tertiary gravels have yielded moderate amounts of gold. The gold-quartz veins are narrow, but the ore often is rich. The ore contains abundant sulfides including pyrite, galena, chalcopyrite, and sphalerite.

Meadow Valley

Location. This district is in west-central Plumas County about eight miles west of Quincy. It includes the Edmanton, Buck's Lake, Spanish Peak and Spanish Ranch areas. The Quincy district lies just to the east and the Granite Basin district to the southwest. The district was mined from the gold rush days through the early 1900s and has been prospected since.

Geology. The east portion of the district is underlain by quartzite, slate, schist, limestone, amphibolite, and serpentine. Meadow Valley is covered by Pleistocene lake beds. Granite lies to the west. There are a number of scattered patches of auriferous Tertiary gravel, in the vicinity of and northwest of Spanish Ranch, which were mined by hydraulicking and drifting. The Pleistocene lake gravels also yielded some gold. The only source of lode gold was the Diadem mine.

Mooreville Ridge

See text under Butte County, p. 18.

Mines. Plumas County: American House hydraulic, Browns Hill, Davis Point hydraulic, Fall River, Sanborn, Walters.

Quincy

Location. This district is in central Plumas County in the general vicinity of Quincy, the county seat. It includes the Elizabethtown and Butterfly Valley areas. The Meadow Valley district lies just to the west and the Sawpit Flat district to the south. The district was first mined during the gold rush, and there has been intermittent prospecting and development work ever since.

Geology. The principal rocks that underlie the district are slate, mica schist, and quartzite. Greenstone lies to the northeast and serpentine to the southwest. American and Thompson Valleys are underlain by Recent and Pleistocene alluvium. A few isolated peaks in the area are capped by basalt.

Ore Deposits. Gold-quartz veins occur principally in slate and mica schist; some are as thick as 15 feet. The veins may be massive or consist of numerous parallel stringers. The ore contains free gold and varying amounts of sulfides, chiefly pyrite. Although some of the veins have been developed for horizontal distances of several thousand feet, none has been worked to depths of greater than a few hundred feet. There are a few small Tertiary channel gravel deposits to the south. The Recent and Pleistocene valley alluvium is gold-bearing in places.

Mines. Placer: Bushman, Carr, Cascade, Elizabethtown Flat, Imperial, Manhattan, Mill Creek, Newton Cons., Newton Flat, Riverdale.

Rich Bar

Rich Bar is in western Plumas County near the junction of the North Fork and the East Branch of the North Fork of the Feather River. During the gold rush this was an extremely rich placer-mining district; Rich Bar alone is credited with an output of $9 million. Later the river was mined by Chinese. The river here goes around a series of sharp bends, a course that has resulted in the formation of wide gravel bars. There are several narrow gold-quartz veins in the area.

Sawpit Flat

Location. This is an extensive gold-bearing region in southern Plumas County. It is contiguous with the Quincy district to the north and the Gibsonville district in Sierra County to the south. It includes the Last Chance, Sawmill Flat, Monitor Flat, Onion Valley, Harrison Flat, Blue Nose Mountain, and Nelson Point areas. The district was named in 1850 for a pit that was dug for the use of a whipsaw. The region was extensively mined during the early days and has been intermittently prospected since. In recent years there has been some placer mining at Monitor Flat. Skin divers have prospected in the Middle Fork of the Feather River.

Geology. The east portion of the district is underlain by slate, schist, and quartzite of the Calaveras Formation (Carboniferous to Permian). Also there are several limestone lenses. To the west the district is underlain by serpentine and amphibolite. The ridges are capped by Tertiary andesite and basalt.

Ore Deposits. The Tertiary gravels are largely quartz-rich, and in places, rich in gold. At Sawpit Flat the gravels are part of the Richmond Hill-Onion Hill channel, while those to the east at Bunker Hill and Blue Nose Mountain are in the northeast end of the famous La Porte channel. A number of gold-quartz veins are in the district, some of which are part of a vein system that extends along the contact zone between serpentine and schist and slate.

Mines. Placer: Boulder West, Bunker Hill, Fordham, Golden Gate, H & G, Kelly, King Solomon, Mayflower, Morning Star Cons., Nelson Creek, Red Slide, Richmond Hill, Rio Vista, Smith, Turkeytown, Union Hill, Zumwalt.

Spring Garden

A number of small lode mines and prospects occur in the general area of Spring Garden and Argentine Rock in south-central Plumas County. A patch of Tertiary auriferous gravel was mined by hydraulicking years ago and the area has been intermittently prospected in recent years. The country rock is greenstone, slate, and quartzite that is overlain to the south by andesite.

Sweet Oil

The Sweet Oil "diggings" in Southwestern Plumas County about eight miles north of La Porte were mined by hydraulicking years ago. The gravel deposits are believed to be located on a branch of the Tertiary La Porte channel. Bedrock is slate, and to the south the gravels are capped by andesite and basalt.

Taylorsville

Location. This district is part of the Crescent Mills-Taylorsville-Genesee gold belt of east-central Plumas County. It has not been as productive as the other two districts in this belt. The general region was first mined during the gold rush, and there has been intermittent prospecting and development work ever since. It was named for J. T. Taylor, who built a mill and hotel there in 1852.

Geology. The Taylorsville area is underlain by a series of northwest-trending belts of Paleozoic and Mesozoic metamorphic rocks, serpentine, and granodiorite. The gold-bearing quartz veins are narrow and strike in a northwest direction. The veins usually occur in and near the granodiorite. The ore contains free gold and varying amounts of pyrite and chalcopyrite.

Mines. Buster, California, Deadman, Iron Dike, King Solomon (placer), Pettinger, Premium $180,000.

• Plumas County Placer Mines

Albion Claim. Owner, Edman Estate, Quincy. Location: Edmanton Mining District, Sec. 32, T. 24 N., R. 8 E., 12 miles southwest of Quincy via Meadow Valley. Elevation 3500'.

Very little work has been done on this property, and it has practically been abandoned.

Bonnie Mine (Eclipse). Owner, Bonnie Hydraulic Mining Company, care W. C. Moran, Berkeley, California. Location: Sawpit Flat Mining District, Sec. 3, T. 22 N., R. 10 E., 25 miles southeast of Quincy to Eclipse. Elevation 6500'.

The property contains 320 acres characterized by lava-capped hills and steeps cañons. It was formerly worked as a hydraulic and drift mine, and in 1913 there were two leasers just starting to drift.

The gold is found in Recent (Pleistocene) and glacial gravels and is probably derived from the quartz veins occurring in the slates on the eastern slope of Pilot Peak.

Water is obtained from Poorman's Creek, a tributary to Nelson Creek and North Fork Feather River.

Boulder West Mine. Owner, Frank Chapman. Location: Sawpit Flat Mining District, Sec. 24, T. 23 N., R. 9 E., 8 miles due south of Quincy, with a total distance via Nelson Point to ridge above mine of 20 miles, then 1 mile of trail. Elevation 4500'.

The holdings consist of one location 20 acres in extent, covering the creek for 1500'. Steep cañons are characteristic of the country.

Surface sluicing only has been done, the values being found in shallow present creek gravels on a bedrock of serpentine. Gold probably derived from quartz and ancient channels above.

Water is obtained from Washington Creek and Middle Fork.

Adjoining mines are the Oversight quartz mine and Baindridge (now Golden Gate).

China Gulch Mine. Owner, M. A. Chaplin, Bucks Ranch. Location: Granite Basin Mining District, Sec. 34, T. 23 N., R. 7 E., 5 miles south of Bucks Ranch, and 30 miles from Quincy. Elevation: 5600'.

The property consists of one claim of 20 acres. It is on the north side of the Middle Fork Feather River, and has been worked for a year by Chinese. One man is sluicing.

Colombo Claim. Owner, G. Giamboni and Company, Oakland. Location: Johnsville Mining District, Sec. 14, T. 22 N., R. 11 E., 1 miles west of Johnsville, then 7 miles to Blairsden. Elevation 6000'.

This property consists of a 20-acre location on glacial gravel. Some work was done on the claim in 1913.

Denning Mine. Owner, Lee F. Denning. Location: Granite Basin Mining District, Sec. 1, T. 22 N., R. 7 E., 7 miles south of Buck's Ranch by trail; thence it is 17 miles northeast to Quincy, by road.

Property is situated on a branch of Willow Creek, one mile from the North Fork Feather River. Owner working alone.

English Bar Group. Owners, Pauly Brothers, Nelson Point, and Judge Moncure, Quincy. Location: Sawpit Flat Mining District, Sec. 11, T. 23 N., R. 10 E., 2 miles northeast of Nelson Point. Elevation 4050'.

At the present time these claims are under option and are being prospected by a company for dredging purposes.

Fenton Group. Owners, R. L. Fenton and Wm. Dushman, Capay, Yolo County. Location: Granite Basin Mining District, Sec. 8, T. 23 N., R. 7 E., 19 miles southwest of Quincy, via Buck's

Ranch. Elevation: 5300'.

Property consists of two claims, the Grizzly Gulch and Lava Gulch, both located in 1913. Assessment work only has been done, pending the securing of a right from the Forestry Bureau to run three-fourths mile of ditch from Lava Gulch.

Flannigan Mine. Owner, Wm. Flannigan, Belden. Location: Butte Valley Mining District, Sec. 28, T. 26 N., R. 7 E.

This property is situated on Soda Ravine, tributary to Mosquito Creek. The creek gravel carries smooth coarse gold and rough quartz gold.

Forest Grove Mine. Owner, Jerry Curtiss, Twain. Location: Johnsville Mining District, Sec. 9, T. 22 N., R. 11 E., 5 miles west of Johnsville, 10 miles southwest of Blairsden by trail. Elevation 6000'.

This property comprises 20 acres, with a length of 2000' along a channel of bench gravel about 1000' above the present creek bed.

It was located in 1884 by B. L. Jones, who took out $7000 by hydraulicking. Curtiss bought it in 1891 and has taken out $3000 by sluicing. One man was working all the time in 1914, but most of the available ground had been worked out.

The deposit is bench gravel, which may have come from the country rock of slate and porphyry. It averages 8' in depth for a width of 75', and has been worked for a length of 2000'.

Four hundred inches of water with an 80' head is obtained with a one and one-half mile ditch from a creek on the north side of Eureka Peak.

The West Elizabeth is an adjoining mine.

Fortuna Mine (Diana). Owner, W. Schield and Company, La Porte. Location: La Porte Mining District, Secs. 31 and 32, T. 22 N., R. 9 E., distant 6 miles from La Porte and 35 miles from Quincy. Elevation 5000'.

The property consists of 150 acres of unpatented land extending three-fourths of a mile along the present stream channel.

The surface gravels have been worked at intervals since early days. It was worked by C. E. Gillette, relocated by Quigley, and in part was purchased by the present owners.

Development consists of two tunnels 150' and 145' in length, under the lava, and a 22' shaft put down 150' beyond the end of the 145' tunnel.

The deposit represents present river gravels, eroded from rocks of the bedrock series and lava gravel, running east. Bedrock has not been reached, as too much water was encountered in the shaft. The deposit is capped with andesite and basalt. The gravel is fine, but has some large boulders in it. A portion lying on 'hardpan' to a depth of 7' is said to have averaged 16¢ to 24¢ per yard. No test pits or drilling have been done, and until some work has been done, the value, depth, etc., are problematical. Surface gravels can only be worked profitably by dredge or dragline scraper.

Garfield Claim (Garibaldi). Owner, P. Laurenzi, Sonoma City. Location: Johnsville Mining District, Sec. 3, T. 22 N., R. 11 E., 4 miles northwest of Johnsville. Elevation 5500'.

This property consists of 40 acres. Assessment work was done in 1913.

Gard and Orr (see Sierra County).

General Harrison Mine (Standard Mining Company). Owner, Chris Radcovich, Johnsville; C. A. Macomber, secretary, San Anselmo. One thousand acre location, dam and two-mile ditch. Location: Johnsville Mining District, Sec. 19, T. 22 N., R. 11 E., 9 miles west of Johnsville. Blairsden is 14 miles northeast by road. Elevation 5000'.

The property embraces 40 acres, covering one-half mile along Nelson Creek, held by the present owner since 1882. The mine consists of present creek gravels on a bedrock of slate. It is now worked out. The Queen group adjoins.

Greenblower Group. Owner, H. J. Greenblower. Location: Granite Basin Mining District, Secs. 30 and 31, T. 23 N., R. 8 W., 6 miles southeast of Bucks Ranch, by trail, thence 17 miles northeast to Quincy by road.

This property consists of a number of claims on the south slope of Mt. Ararat, two miles north of the Feather River. Owner working alone.

Hellas Mine. Owner, Edman Estate, Quincy. Location: Edmanton Mining District, Sec. 33, T. 24 N., R. 8 E., 2 miles southeast of Meadow Valley, thence 9 miles by road east to Quincy. Elevation 3500'.

This property covers creek gravels, but it has practically been abandoned.

Jones Placer Mine (Weaver). Owner, David Jones, Waldo, Yuba County, California. Location: Edmanton Mining District, Sec. 7, T. 23 N., R. 8 E., 2 miles east of Bucks Ranch by trail, thence 17 miles northeast to Quincy by road. Elevation 5200'.

One man, J. C. Kaiser, has been working on the claim, ground sluicing, for the past four years.

Joseph Group (Emerson and McMullin mine). Owner, A. Joseph, Quincy. Bonded to R. C. Jackson, Folsom and Twenty-second streets, San Francisco. Location: Genesee Valley Mining District, Secs. 2 and 11, T. 24 N., R. 11 E., 7 miles south of Genesee by trail; 22 miles northwest of Portola by road. Elevation 5280'.

King Solomon Group. Owners, Lucky Strike Mining Company, H. O. Howard, Lovelock, Nevada. Location: Taylorsville Mining District, Sec. 30, T. 25 N., R. 11 E., 6 miles southeast of Taylorsville, thence 12 miles southwesterly by road to Kiddie.

The King Solomon Group of six placer claims was acquired early in 1915 by the above company. The company took a bond on the group in 1912 and has performed sufficient work to demonstrate the presence of rich gravel. The channel has been traced for a mile, and at places is 300' wide. Most of the work consists of shafts and open cuts, but preparations are being made for more thorough development.

Little Star Mine. Owner, S. A. Pezzola, Johnsville. Location: Johnsville Mining District, Sec. 24, T. 22 N., R. 11 E., near border of Johnsville, 9 miles west of Blairsden by road. Elevation 5200'.

This property comprises 40 acres of locations in the creek bed just above Johnsville. It has a length along the channel of 3000'.

The claims were relocated by the present owner in 1909. About 20 acres of the property was worked in early days. Assessment work only being done at present.

The deposit consists of free glacial and creek gravels, with some boulders. They are the surface gravel of Jamison Creek, the gold being derived from the erosion of the Plumas Eureka and Jamison veins. The channel courses north.

The Jamison placer and Plumas Eureka quartz mines adjoin.

McFarlane Mine. Owner, Vernon Hammet. Location: Edmanton Mining District, Sec. 7, T. 23 N., R. 8 E., 3 miles east of Bucks Ranch by trail, thence 17 miles northeast to Quincy by road.

This mine is idle at present, owner working on the Star Plumas claim.

Minerva Bar Claim. Location: Sawpit Flat Mining District, Sec. 18, T. 23 N., R. 10 E., 3 miles west of Nelson Point. Elevation 3800'.

Formerly owned by Minerva Bar Mining Company. Last work done by Scott Beaser in 1911.

Nelson Creek Claim. Owners, E. C. Fish, Oroville; C. M. Root, Nelson Point. Location: Sawpit Flat Mining District, Sec. 22, T. 23 N., R. 10 E., 1 mile south of Nelson Point, thence 11 miles northwest to Quincy by road. Elevation 4000'.

Assessment work only being done.

New York Mine. Owner, N. H. Fries. Location: Granite Basin Mining District, Sec. 15, T. 23 N., R. 7 E., 2 miles south of Buck's Ranch, thence 17 miles northeast to Quincy by road.

This property consist of 40 acres, with ground owned by Gold Mountain Hydraulic and Dredging Company. Owner working alone.

Plumas Investment Company. Owner, Plumas Investment Company, W. P. Hammon, 433 California Street, San Francisco. Location: Spanish Ranch Mining District. Elevation 3700'-3900'.

The company's holdings comprise three groups: The Mountain House No. 59, four miles northwest of Spanish Ranch in Sec. 31, T. 25 N., R. 8 E.; the Pine Leaf No. 91, one mile north of Spanish Ranch in Sec. 11, T. 24 N., R. 8 E., and the Silver Star group containing six hundred acres in Sec. 15, T. 24 N., R. 8 E., about one mile west of Spanish Ranch. The latter group takes in Spanish Creek Valley, west of town. No work is being done on any of these claims.

Quartz Ravine Claim. Owner, H. Smith. Location: Granite Basin Mining District, Sec. __, T. 22 N., R. 7 E., 7 miles by trail south of Buck's Ranch, thence 17 miles northeast to Quincy by road.

This deposit consists of creek gravels located south of China Gulch claim.

Rock Rim Mine. Location: Granite Basin Mining District, Sec. 2, T. 22 N., R. 7 E., 7 miles south of Bucks Ranch by trail, thence 17 miles northeast to Quincy by road.

Situated on Willow Creek, one mile from Middle Fork Feather River, near the China Gulch mine. Owner is doing assessment work alone.

Woods Ravine Mine. Owner, R. Gifford, Spanish Ranch. Location: Spanish Ranch Mining District, Sec. 7, T. 24 N., R. 9 E., 1-1/2 miles northwest of Spanish Ranch, 9 miles to Quincy by road. Elevation 3800'.

This property consists of 40 acres in a ravine fed from the Badger and Gopher Hill mines and the tailings are reworked. No work has been done in the last few years.

Dredging

Gold Mountain Hydraulic and Dredging Company.

Mountain Meadow Dredging Company. Owner, same, Chicago; E. E. Sirling, in charge. Location: Crescent Mills Mining District, 8 miles from Susanville.

This property consists of 3000 acres located by Lloyd Baker of Chicago. It was reported that a dredge was to be installed during 1915, but the project did not materialize. [9]

• Innis Dredging Company of Nevada City moved its dragline dredge to Lights Creek in the Lights Canyon district and resumed operations on August 4, 1940. In 1941 this operation was continued from January 1 to September 22, and the yield from 250,000 cubic yards of gravel was 1653 ounces of gold and 130 ounces of silver. [1]

SACRAMENTO COUNTY

Folsom

Location. This district is in northeastern Sacramento County. It was mainly a dredging field that extended from the town of Folsom southwest along the south side of the American River to Fair Oaks, south through the town of Natoma to Nimbus and then west to the east border of what is now Mather Air Force Base. The dredged area is approximately 10 miles long in a southwest direction and up to seven miles wide. The Folsom district has also been known as the American River district.

History. The region around Folsom and Mormon Bar was extensively placer-mined during the gold rush, with minor lode mining. The area was originally settled in 1849 and first known as Negro Bar. The present town was laid out in 1855 by T. D. Judah for the Sacramento Valley Railroad and named for Captain J. L. Folsom, quartermaster of Stevenson's Regiment. Numerous Chinese worked the region from the 1860s through the 1890s. A primitive grab-dredger was active at Natoma in 1894. Bucket-line dredging began at Folsom in 1898 and soon became a major industry. Most of the dredging companies were merged into Natomas Consolidated of California in 1908. This firm, later known as the Natomas Company, was the principal operator in the district. The company designed and built its own dredges at extensive shops in the town of Natoma. In 1916, 11 active dredges yielded more than $2 million worth of gold. From 1927 to 1952, several other operators joined Natomas in dredging the district.

Dredging operations were curtailed during World War II but were resumed on a major scale shortly afterward. However, increasing costs, the depletion of dredging ground, and changing land values caused the dredging operations to be gradually curtailed. By 1960 there was only a single active dredge, and this was shut down in February 1962. Large portions of the dredged-over areas are now occupied by defense industries, such as the Aerojet-General Corporation and Douglas Aircraft Company plants, and by housing tracts. Folsom, one of the largest dredging fields in California, has a total output estimated at $125 million. Approximately one billion cubic yards of gravel were dredged by the Natomas Company.

Geology. Recent stream gravels lie in and adjacent to the present American River. To the south are sand and gravel deposits of the Victor Formation (Pleistocene) and silt, sand and some gravel of the Laguna Formation (Plio-Pleistocene). These are underlain by andesite of the Mehrten Formation (Pliocene). The paying gravels are either in or along the American River and near the lower contacts of the Laguna and Mehrten Formations. Digging depths ranged from 30 to 110 feet and recoveries from 10¢ to 20¢ per yard with gold valued at $35 per ounce. Minor amounts of platinum were recovered. There are a few narrow gold-quartz veins in greenstone east of the town of Folsom.

Dredging Operations. Capitol Dredging Co., 1927-52, four bucket-lines; General Dredging Co., 1938-51, three draglines; Gold Hill Dredging Co., 1933-37, one bucket-line; Lancha Plana Gold Dredging Co., 1940-49, one bucket-line. Natomas Co., 1909-62, fifteen bucket-lines; consolidated by Natomas Co. in 1908 were: Colorado Pacific Gold Dredging Co., El Dorado Gold Dredging Co., Folsom Development Co., Natoma Development Co., Syndicate Gold Dredging Co., Wilkes-Barre Dredging Co. (1916); later Natomas Co. production records, 1959, 7,894,592 cu. yds., 10.19¢/yd. yield, 9.56¢ cost per yard and 1958, 9.15¢/yd. yield, 8.73¢ cost per yard.

Michigan Bar

Location and History. This is a placer-mining district in eastern Sacramento County in the vicinity of the old town of Michigan Bar. It extends to the west and south and includes the Sloughhouse area. Included are dredging fields in and near the Cosumnes River. The district was hydraulicked extensively in the 1850s and 1860s. Later it was worked by small scale methods by many Chinese miners. The Cosumnes River and some of the older bench gravels were dredged in the 1930s and early 1940s. Also some drift mining was done. Fire clay is mined here now. The total gold production is unknown, but it has been estimated to be at least 1,700,000 ounces.

Geology. Eocene gravels interbedded with several layers of clay and thin sands are distributed over the eastern part of the district in the vicinity of Michigan Bar. The gravels are coarse, well rounded, have a sandy matrix, and usually are not too well-cemented. The dredged area is several miles long.

• Sacramento County Dredging Operations

In 1908 the production of gold from the various mines in Sacramento County amounted to $1,166,055. Of this $409 came from the hydraulic mines, $52,365 from drift mines, $3,925 from surface placers, and $1,109,196 from dredging operations.

The total known dredgeable area in Sacramento County in 1908, including that which has been dredged, comprised 6,052.66 acres.

All the dredge mining in Sacramento County is carried on along the American River in what is known as the American River or Folsom district. This district includes in its boundaries a part of the town of Folsom, and extends 9 miles from Folsom to Cornell, a place on the Sacramento and Placerville Railroad, 11 miles by rail from Sacramento City.

The total area of the district comprises 12,522.58 acres, of which about 6,052.66 acres have so far proven dredgeable. All of this land, with the exception of 1,326 acres on the Sacramento, Sailor, and Mississippi bars, is located on the south side of the American River, extending in places 2-1/2 miles inland. All of the land in the district, with the exception of 431 acres, is owned by the Natomas Consolidated of California, the larger of the two companies operating in the field.

The production of gold from dredging operations in the American River district from 1899 to December 31, 1908, amounted to $3,920,231.

The American River District in 1909 and during the first half of 1910.

On January 1, 1909, the Natomas Consolidated of California took over the holdings and equipment of three companies operating in the district. These were: The Folsom Development Company, the Natoma Development Company, and the El Dorado Gold Dredging Company. The Natomas Consolidated also absorbed the holdings of the Natoma Vineyard Company. [2]

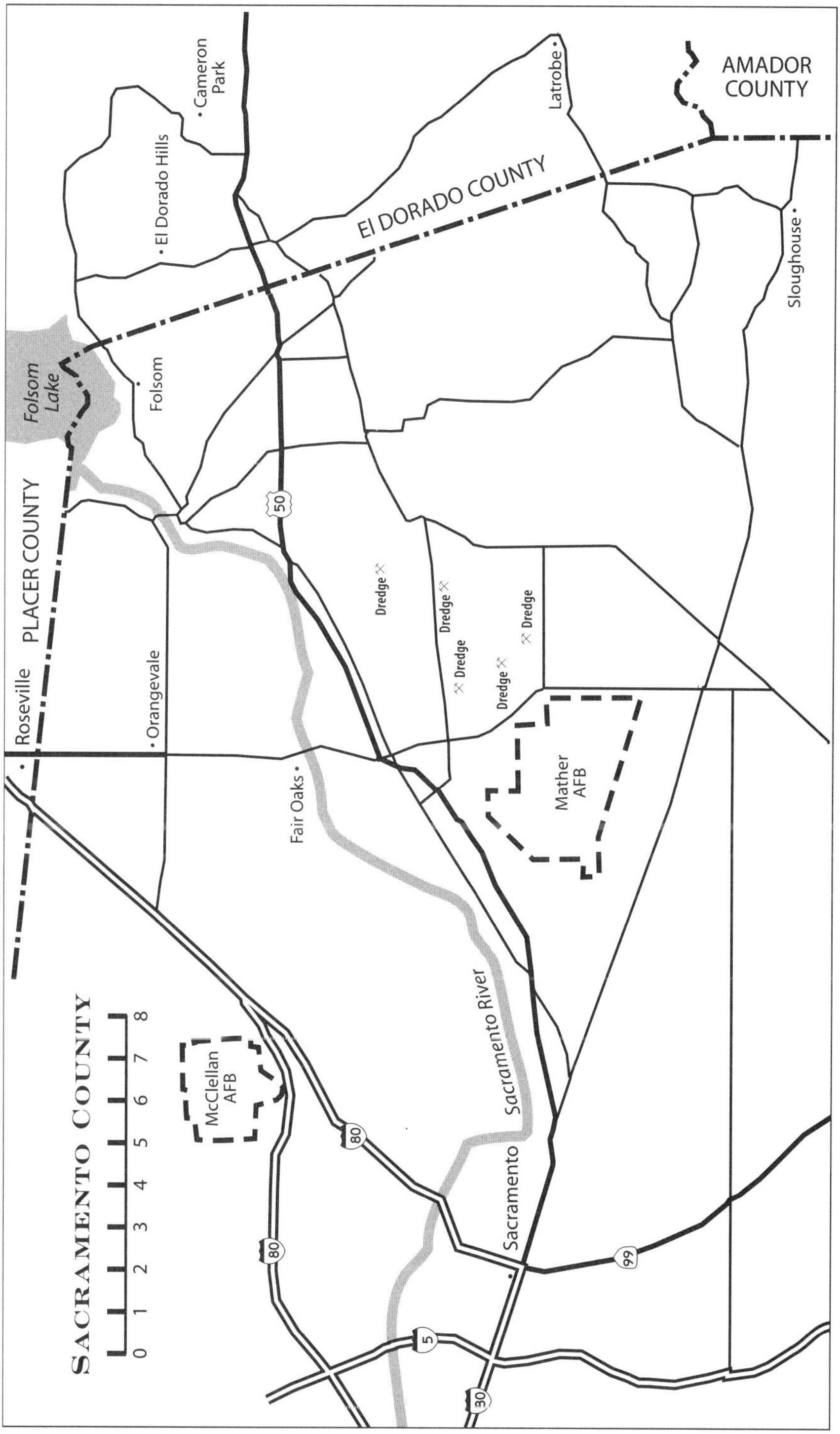
Sacramento County
0 1 2 3 4 5 6 7 8
Cameron Park
El Dorado Hills
Latrobe
AMADOR COUNTY
EL DORADO COUNTY
Sloughouse
Folsom Lake
Folsom
PLACER COUNTY
Roseville
Orangevale
Fair Oaks
Dredge
Dredge
Dredge
Dredge
Dredge
Mather AFB
McClellan AFB
Sacramento River
Sacramento
50
80
80
99
5
30

SIERRA COUNTY

Alleghany

Location. Alleghany is in southwestern Sierra County. This district is in a belt of gold mineralization that extends from Goodyear's Bar, south and southeast through Forest, Alleghany, Chip's Flat, and Minnesota. This gold-bearing belt continues south to the Washington district in Nevada County. The Downieville and American Hill districts are to the east, and the Pike district is to the west.

History. The streams in the area were placer-mined soon after the beginning of the gold rush, and the Forest diggings were discovered in the summer of 1852 by some sailors. Some of these sailors were "Kanakas" or Hawaiians who also had deserted their ships in San Francisco. Forest, first known as Brownsville and then Elizaville, got its present name in 1853. The Bald Mountain and other drift mines were highly productive from then until around 1885. Hydraulic mining was done at Minnesota and Chip's Flat during these years. The town of Alleghany was named for Alleghany, Pennsylvania.

Quartz mining was reported to have begun in the district in 1853 at the German Bar and Ireland mines. Although the quartz mines were moderately productive until the 1870s, drift mining was the principal source of gold then. The rediscovery of the Tightner vein in 1904 by H. F. Johnson (erroneously given as 1907 in many reports) led to the revival of lode mining, which continued until 1965.

Alleghany was the only town in California after World War II where gold mining was the principal segment of the economy. After 1960, production from the district, which had been averaging more than $500,000 per year, decreased greatly as more and more mining operations were curtailed. By 1963, the output was less than $100,000 per year. The Sixteen-to-One mine, the largest gold source in the district, curtailed normal operations late in 1962, and the Brush Creek mine, the second largest operation, was shut down in 1964. At the end of 1965 the Sixteen-to-One mine was completely shut down, ending an operation that had lasted more than 60 years. Intermittent operations have continued at several mines, such as the Kate Hardy, Oriental, El Dorado-Plumbago, and Mugwump mines. Several of the mines received Federal exploration loans. Skin divers are active in the streams of the area.

Alleghany was the most famous high-grade gold mining district in California. The value of the total output is unknown, estimated at $50 million. Much of this production was from small but spectacularly rich ore bodies.

Geology. The district is underlain by north- and northwest-trending beds of metamorphic rocks of the Calaveras Formation (Carboniferous to Permian), serpentine, and greenstone. In the vicinity of Alleghany and Forest this formation has been divided into six units: Blue Canyon Slate, Tightner Formation (chiefly amphibolite and chlorite schist), Kanka Formation (conglomerate, chert, and slate), Relief Quartzite, Cape Horn Slate, and the Delhi Formation (phyllite and slate). These rocks have been invaded by many basic and ultra-basic intrusions; the ultra-basic rocks have been largely serpentinized. Mariposite-bearing rock, locally known as "bluejay," is commonly adjacent to the serpentine. Also present are fine to medium-grained dioritic dikes. The higher ridges are capped by andesite and basalt, which in places overlies auriferous Tertiary channel gravels.

Ore deposits. The gold-quartz veins strike in a northerly direction, dip either east or west, and usually range from two to five feet in thickness. They occupy minor reverse faults, and occur in all of the rocks of the Calaveras Formation, and in the greenstone. The largest number of mines are in amphibolites of the Tightner Formation. The most characteristic features of the ore deposits are the extreme richness, erratic distribution and small size of the ore shoots. They range from small masses of gold and quartz yielding a few hundred dollars to ore bodies that have yielded hundreds of thousands of dollars. One ore body at the Sixteen-to-One mine, which had a pitch length of 40 feet, contained nearly $1 million, while another at the Oriental mine about 14 feet long yielded $734,000.

The gold occurs in the native state commonly with arsenopyrite but only small amounts of other sulfides. In a few places pyrite is abundant. The numerous serpentine bodies and associated mariposite rock are structurally important in the localization of the ore bodies. The quartz veins tend to fray or bend near serpentine, and it is in these frayed or bend portions of the veins that the high-grade ore bodies are often found. Highgrade ore also is found in vein junctions or in sheared portions of the veins.

Channel gravels. A major tributary of the Tertiary Yuba River extended south from Rock Creek through Forest and Alleghany and then southeast through Chip's Flat and Minnesota to Moore's Flat in Nevada County. This is commonly known as the "Great Blue Lead" or Forest channel. It was uniformly rich except where cut by later channels. The largest gold producers were the Ruby, Live Yankee, and Bald Mountain drift mines, where many coarse nuggets were recovered. During the late 1930s a number of fist-sized gold nuggets were recovered from the Ruby mine. These were displayed for many years in the Sierra County exhibit at the California State Fair in Sacramento.

Mines. Drift: Bald Mountain $3.1 million, Bald Mountain Extension $500,000 to $1 million, Gold Star $250,000+, Highland & Masonic $300,000+, Live Yankee $750,000 to $1 million, Ruby $1 million+.

American Hill

Location. The American Hill district is in southwestern Sierra County about five miles east of the town of Alleghany. It is both a lode and placer district, but the placer deposits have been more important. It includes the Cornish House area.

Geology. The district is chiefly underlain by slate. Serpentine and amphibolite are to the west, and granodiorite is just to the east. Extensive gravel deposits are part of a tributary to the Forest channel of the Tertiary Yuba River, which extends in a southwest direction through the district. The northern part of the district is covered by andesite. Dioritic dikes often are associated with the gold-quartz veins.

Ore deposits. The gravels are quartzitic and contain coarse gold. One of the gravel deposits is as much as 300 feet thick and covered with clay and sand. The quartz veins are lenticular and occur either in the slate or near the slate-granodiorite contact. The ore bodies contain free gold with pyrite, arsenopyrite, and galena. Sometimes carbon is found in cavities in the quartz. Some high-grade pockets have been found, but most of ore averages less than 1/2 ounce per ton.

Mines. Placer: American Hill, Bear Creek, Excelsior, Mable Mertz, Yellow Jacket.

Brandy City

Location. This district is along the extreme west margin of Sierra County about 12 miles west of Downieville and 10 miles by road north of Camptonville. It is on the ridge between Canyon

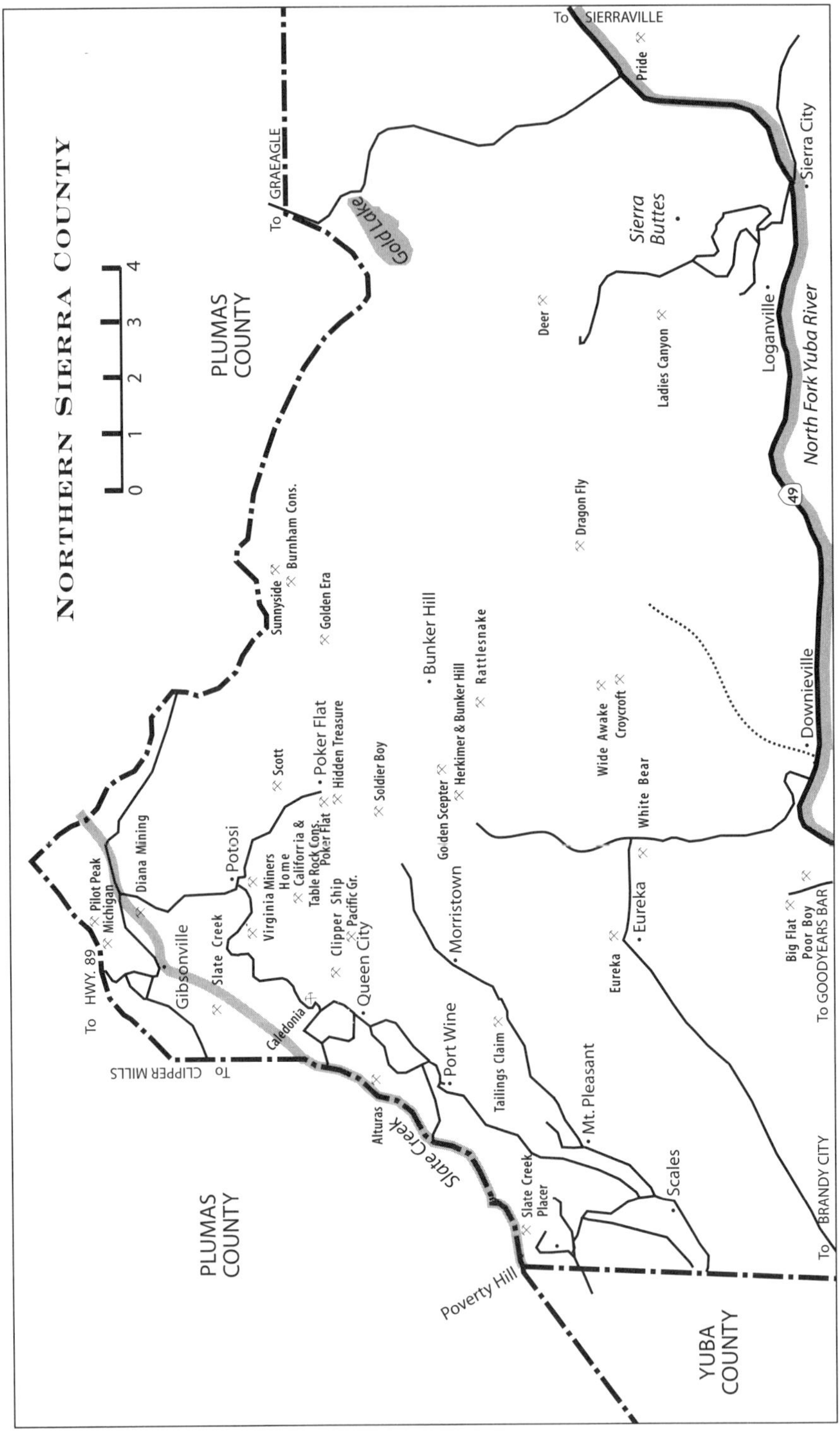
NORTHERN SIERRA COUNTY
0 1 2 3 4
PLUMAS COUNTY
PLUMAS COUNTY
YUBA COUNTY
To GRAEAGLE
To SIERRAVILLE
To HWY. 89
To CLIPPER MILLS
To GOODYEARS BAR
To BRANDY CITY
Gold Lake
Sierra Buttes
Sierra City
Loganville
North Fork Yuba River
49
Pride
Deer
Ladies Canyon
Dragon Fly
Sunnyside
Burnham Cons.
Golden Era
Bunker Hill
Rattlesnake
Herkimer & Bunker Hill
Golden Scepter
Wide Awake
Croycroft
White Bear
Downieville
Scott
Poker Flat
Hidden Treasure
Soldier Boy
Potosi
Virginia Miners Home
California & Table Rock Cons.
Poker Flat
Clipper Ship
Pacific Gr.
Queen City
Diana Mining
Pilot Peak
Michigan
Gibsonville
Slate Creek
Caledonia
Morristown
Port Wine
Tailings Claim
Eureka
Eureka
Big Flat
Poor Boy
Mt. Pleasant
Alturas
Slate Creek
Slate Creek Placer
Scales
Poverty Hill

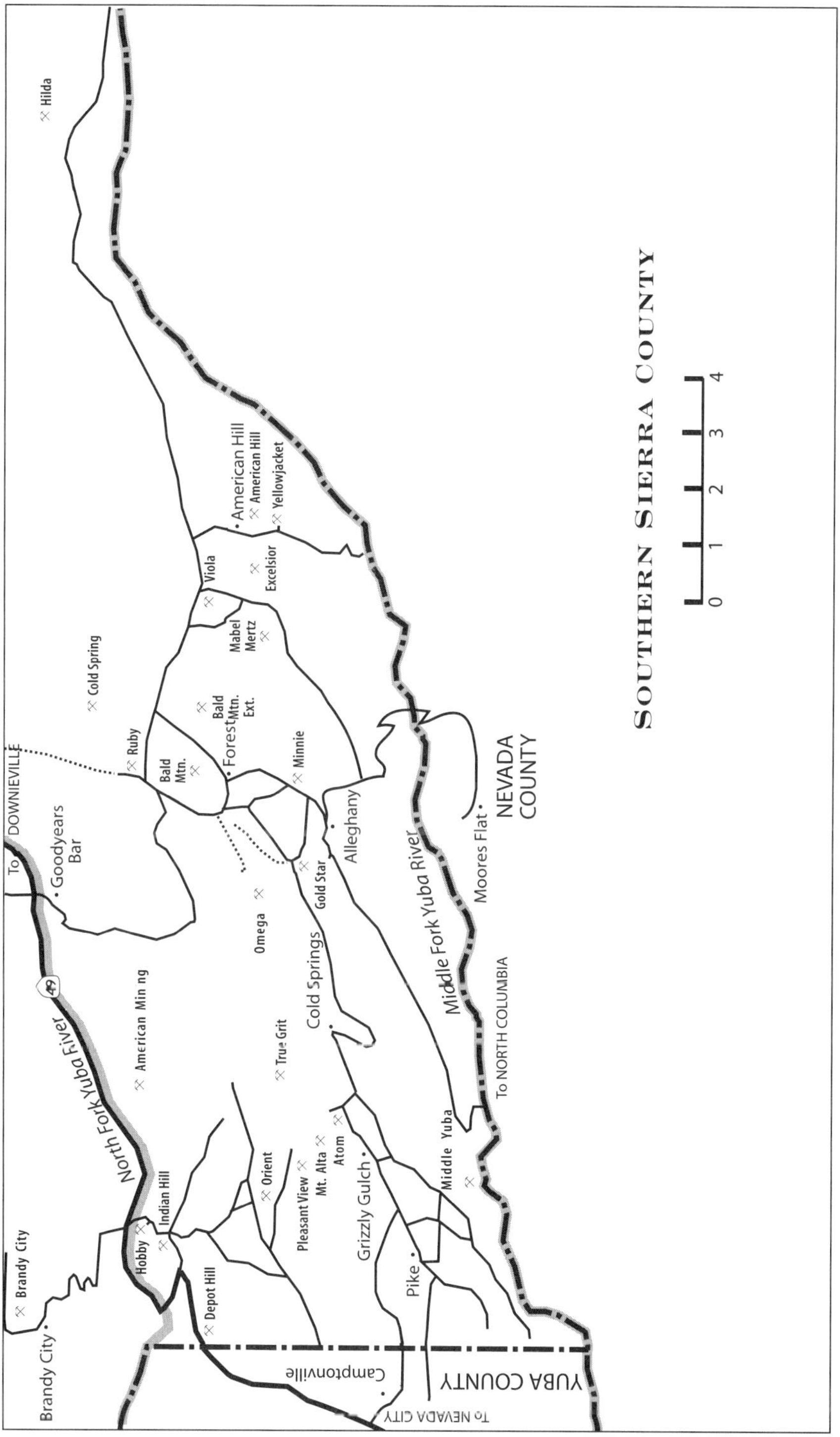
SOUTHERN SIERRA COUNTY
0 1 2 3 4
Hilda
Cold Spring
To DOWNIEVILLE
Goodyears Bar
Ruby
Bald Mtn.
Forest
Bald Mtn. Ext.
Mabel
Mertz
Viola
Excelsior
American Hill
American Hill
Yellowjacket
Minnie
Alleghany
Gold Star
Omega
Cold Springs
Middle Fork Yuba River
Moores Flat
NEVADA COUNTY
To NORTH COLUMBIA
North Fork Yuba River
49
American Mining
True Grit
Brandy City
Brandy City
Hobby
Indian Hill
Depot Hill
Orient
Pleasant View
Mt. Alta
Atom
Grizzly Gulch
Middle Yuba
Pike
Camptonville
To NEVADA CITY
YUBA COUNTY

Creek on the north and the North Fork of the Yuba River on the south. It is principally a placer-mining district and was extensively hydraulicked from the 1850s until the early 1890s. Some work was done again from the early 1900s to the 1930s. The value of the total production of the district is unknown, but it probably amounts to several million dollars' worth of gold.

Geology. The main channel of the Tertiary North Fork of the Yuba River, which is also known as the La Porte channel, extended in a south-southwest direction from Poverty Hill into this district and thence to Indian Hill. Pay gravel here is 700 to 800 feet wide, up to 200 feet thick, and overlain by 40 to 60 feet of andesite. The upper 130 feet of gravel are mostly pebbles; the lower portion is bouldery and well cemented.

The entire deposit is reported to have averaged 25¢/yard in gold at the old price of gold, and near bedrock it was reported to have contained as much as $2.50/yard in gold. Bedrock is slate and serpentine with amphibolite both to the east and west.

Camptonville

Camptonville, in northeast Yuba County and western Sierra County, adjoins the Pike district on the east. Rich gold discoveries were made here in 1850-51. The town was named in 1854 for Robert Campton, the local blacksmith. The Pelton wheel, long used in mining machinery and electric generators, was invented here by Lester Pelton. The old town is now a tourist attraction. A number of moderate-sized deposits of Tertiary channel gravel were mined by hydraulicking. These include the deposits at Young's Hill, Weed's Point, and Galena Hill, They are part of the Tertiary Yuba River that extended southwest from Indian Hill to this district and then to North San Juan. Bedrock is amphibolite, slate, greenstone, and granodiorite.

Downieville

Location.. This is a lode and placer gold-mining district in west-central Sierra County in the general vicinity of the town of Downieville. It includes the Fir Cap Mountain, Craycroft, China Flat, and Slug Canyon areas and part of the Pliocene Ridge area. The Goodyear's Bar-Allegheny belt lies immediately to the west, the Sierra City district to the east, and the American Hill district to the south.

History. This area was prospected soon after the beginning of the gold rush. Major William Downie and his party arrived here in November, 1849. Soon a town was laid out, which was named for him early in 1850. The Downieville mining district was organized with "claims fixed at 30 feet per man". Many rich strikes were made; one claim 60 feet square yielded $80,000 in six months. At nearby Tin Cup Diggings, three men filled a tin cup with gold each day before quitting. A 25-pound nugget was found in the river upstream from the town in 1850. More than 5000 persons lived here in 1851. After the surface placers were exhausted, the river was mined, and hydraulic and drift mining became important. Mining continued almost steadily until World War II, and intermittent prospecting and skin diving for gold continues.

Geology. The district is underlain predominantly by north-trending beds of phyllite, slate, and quartzite of the Calaveras Formation (Carboniferous to Permian). To the west are greenstone, amphibolite, and serpentine. The higher ridges are capped by Tertiary andesite, which in places overlies rich Tertiary auriferous gravels. There are fairly extensive Recent river and terrace gravels along the Yuba River and its branches.

Ore Deposits. A considerable number of gold-quartz veins occur chiefly in greenstone and slate. The veins range front one to 10 feet in thickness. The ore contains free gold and varying amounts of sulfides.

The milling-grade ore usually averages 1/4 to 1/3 ounce per ton. Some ore shoots had stoping lengths of as much as several hundred feet. Some high-grade ore pockets have been taken from some of the mines. The Tertiary channel gravels are quartzitic, often well-cemented, and in places contain extremely coarse gold.

Mines. Placer: Brown Bear, City of Six, Craycroft, Golden Hub, Kirkpatrick, Klondike, Mott and Mt. Vernon, Monte Carlo, New York, White Bear $200,000+, Wide Awake $100,000+.

Eureka

Location. This is an extensive area of scattered placers and a few lode deposits in northwestern Sierra County about eight miles northwest of Downieville. It includes the "diggings" not only at Eureka but also at Craig's Flat, Morristown, and Saddleback Mountain.

It is surrounded by a number of famous placer-mining districts: Downieville, Poker Flat, Port Wine, Poverty Hill, and Brandy City. The hydraulic mines here were worked on a major scale from the 1850s to the middle 1880s, and then intermittently on a small scale through the 1930s.

Geology. The principal Tertiary channel deposits are at Eureka, Craig's Flat, Morristown, and Monte Cristo, the most extensive being at Eureka. They are part of the Eureka channel, an indistinct branch of the Tertiary North Fork of the Yuba River. As in the other nearby placer-mining districts, the chief values were obtained from the lower quartzitic gravels. Some very coarse nuggets have been found here. Bedrock consists of slate and phyllite and several narrow belts of greenstone and serpentine. Several of the high ridges are capped by andesite. There are a few gold-quartz veins, the most productive having been at the Telegraph mine, which is on a slate-serpentine contact.

Gibsonville

See text under Plumas County, p. 130.

Indian Hill

Indian Hill is in western Sierra County about 10 miles west of Downieville and just south of the Brandy City district. Much of the production here has come from the Indian Hill and Depot Hill hydraulic mines. These mines were extensively worked from the 1850s to the 1880s, and intermittent development work and mining continued through the 1930s. The Depot Hill mine has been prospected recently. The deposits are on the LaPorte-Brandy City Tertiary channel. The lower gravels contain abundant quartz and are as much as 100 feet thick. They are overlain in places by intervolcanic gravels and andesite. Bedrock is granite with amphibolite to the west and slate, schist, and serpentine to the east. There are also a few gold-quartz veins in the district.

Pike

Location and History. This district is in the southwest corner of Sierra County at the site of the old town of Pike or Pike City. Some of the early-day miners came from Pike, Missouri. The district, 12 miles northeast of North San Juan and 26 miles northeast of Nevada City, includes the Tippicanoe, Negro Tent, Snowden Hill, and Grizzly Gulch areas. The area was originally mined during the gold rush, and the Alaska mine was worked on a large scale from

1863 to 1916. The Pleasant View hydraulic mine was active in 1962 and 1963.

Geology and Ore Deposits. The district is underlain by metadiabase, serpentine, and irregular bodies of amphibolite. Fine-grained phyllite and quartzite of the Delhi Formation (Carboniferous) lie on the cast side.

Lenticular quartz veins with calcite occur chiefly in the metadiabase. The ore bodies contain free gold and often abundant sulfides, including galena. There are some extensive ore shoots and some high-grade pockets. The gold often is coarse. Several patches of Tertiary quartz-rich gravel are a few acres in extent, some of the gravel capped by andesite.

Mines. Placer: Grizzly Gulch, Mt. Alta, Orient, Pleasant View, Tippicanoe, True Grit.

Poker Flat

Location. This district is in northern Sierra County about 10 miles north of Downieville. It includes the Howland Flat, Table Rock, Deadwood, Mt. Fillmore, Potosi, and Rattlesnake Peak areas. It is mainly a placer-mining district.

History. The streams were first mined during the gold rush. The locality was extremely rich then; in one month gold valued at $700,000 was produced. Hydraulic mining was done on a major scale from the late 1850s through the 1880s. Some lode mining and drifting continued through the early 1900s, and the area was prospected in the 1920s and 1930s. The district was made famous by Bret Harte's tale, *The Outcasts of Poker Flat.* This district has been highly productive, the mines at Howland Flat alone being credited with an output valued at $14 million.

Geology. The northern part of the district is underlain by amphibolite with some serpentine. To the south and east there are slates of the Blue Canyon Formation (Carboniferous). Substantial portions of the area are capped by andesite. Extensive deposits of Tertiary auriferous quartz gravels are part of the Port Wine channel, which extends west and northwest through this district and then west and southwest into the Port Wine district. The lower quartz-rich gravels were also gold-rich. Portions of the channel have been faulted. Some narrow gold-quartz veins occur in amphibolite and slate.

Mines. Placer: Caledonia, California, Clippership, Deadwood, Forest Queen, Gibraltar, Hawkeye, Herkimer and Bunker Hill Manchester, Miners Home, Pacific, Poker Flat, Potosi, Rattlesnake, Scott, Virginia, Tennessee, Winkeye.

Port Wine

Location. The Port Wine district is in northwestern Sierra County about 10 miles northwest of Downieville. The site was named by a party of prospectors who found a keg of port concealed in the bushes. The La Porte district adjoins on the northwest, the Poverty Hill district on the southwest, and the Poker Flat-Rowland Flat district on the northeast. This district includes the "diggings" at Grass Flat, Queen City, and St. Louis. The area was extensively mined by hydraulicking and drifting during and after the gold rush and has been intermittently prospected since.

Geology. The Port Wine channel, a major branch of the La Porte channel, extends southwest through this area. It is roughly parallel to the La Porte channel for some miles and joins it at Scales. The channel is well-defined and several hundred feet wide. The gravels are quartz-rich and in places are covered by "pipe" clay and andesite. Bedrock is slate, quartzite, amphibolite, and greenstone.

Poverty Hill

Location and History. This district is in western Sierra County, 10 miles northwest of Downieville and five miles south of La Porte. The Port Wine district is to the northeast, and the Brandy City district is to the south-southwest. Poverty Hill district includes the Scales and Mount Pleasant areas. The area was mined by hydraulicking and drifting during and after the gold rush. Mining here again during the 1930s and early 1940s included an attempt to work the Poverty Hill pit with a bucket-line dredge.

Geology. The main channel of the Tertiary North Fork of the Yuba River, or La Porte channel, enters this district from the north and continues south and southwest to the Brandy City and Indian Hill districts. The Port Wine channel, a branch of the La Porte channel, enters the area from the northeast at Scales. This smaller branch channel parallels the main channel for some miles. At the Poverty Hill pit, the channel is up to 1500 feet wide and 150 feet deep. The lower "blue" gravels are quartz-rich and cemented in places. There are a number of large boulders. At Scales the channel is similar. Bedrock consists of slate, quartzite, amphibolite and serpentine. Lindgren estimated in 1911 that, at Poverty Hill, 2.25 million yards had been removed and 5 million yards were ultimately available, while at Scales and Mt. Pleasant, 4.05 million yards had been excavated and 60 million yards were ultimately available.

Sierra City

Location. The Sierra City district covers an extensive area in central Sierra County. It includes the Former, Loganville, Church Meadows, Gold Valley, and Sierra City-Sierra Buttes areas. It is located at the south end of the major belt of gold mineralization that extends north-northwest to the Johnsville district in Plumas County.

History. This district was placer-mined soon after the beginning of the gold rush. Many coarse nuggets were recovered at that time. Sierra City was founded in 1850; destroyed by an avalanche in 1852, it was soon rebuilt. Many Indians lived in this district in those days. The famous roistering society E. Clampus Vitus originated in Sierra City. It was reorganized several years ago by the California Historical Society.

The Sierra Buttes mine was opened in 1850, and most of the other important lode mines soon afterward. A number of very rich high-grade surface pockets were discovered, including one at the Four Hills mine that yielded between $250,000 and $500,000. The district was highly productive from about 1870 until 1914. There was some mining activity again during the 1920s and 1930s, and intermittent prospecting and development work has continued until the present time. The value of the total output of the district is unknown, but it is estimated to be at least $30 million.

Geology. A north-west-trending belt of Calaveras (Carboniferous to Permian) slate, schist, and quartzite with limestone lenses runs through the west portion; a quartz porphyry (metarhyolite) belt lies in the central portion, and greenstone and amphibolite schist, to the east. A belt of meta-tuff of the Milton Formation (Jurassic) extends along the east margin. A few serpentine lenses are present. The northeastern area is overlain by glacial moraines. Tertiary andesite caps some of the ridges and rhyolite caps some Eocene auriferous channel gravels.

Ore Deposits. An extensive series of north- to northwest- and occasionally northeast-trending quartz veins range from a few feet to as much as 40 feet in thickness. The veins occur in all of the metamorphic rocks. The ore bodies contain free gold, pyrite, and minor galena and chalcopyrite. The ore shoots often were large (up to 300 feet long and several thousand feet deep) and usually averaged 1/3 to 1/2 ounce of gold per ton. Much high-grade ore was taken from the district. Several fairly extensive magnetite deposits are found in the north end of the district.

Mines. Placer: Hilda, Ladies Canyon $500,000, Pride.

• Sierra County Gold - Placer Mines (Surficial).

Alturas Mine. Owner, Maschella, San Francisco. Location: Port Wine Mining District; 1 mile west of Port Wine, thence 30 miles northerly, by road, to Quincy. Elevation 4000'.

This property has been worked for the past thirty years. Ten to fifteen men employed in 1914. Gravel is elevated from the creek and sluiced.

American Mining and Development Company. Owners, American Mining and Development Company. Location: Indian Hill Mining District; Secs. 14 and 15, T. 19 N. R. 9 E., 2 miles northeast of Indian Hill, thence 40 miles by road southerly to Nevada City.

These holdings comprise 175 acres of patented land.

Antlered Crest Mine. Owner, Geo. Sanborn, La Porte. Location: Sierra City Mining District. Sierra City is 56 miles by road, northeast of Nevada City and 20 miles, by road, south of Blairsden.

A hard bedrock tunnel was being run in 1914 and was in 430'.

Big Flat Claim. Owner, W. W. Casserly, Goodyear's Bar. Location: Downieville Mining District, near Poker Flat on east side of Port Wine Ridge. Poker Flat is 33 miles, by road, south of Quincy.

Bon Ton Consolidated Group. Owner, same. Location: Poker Flat Mining District; Sec. 28, T. 20 N., R. 12 E., 56 miles, by road, northeast of Nevada City, 20 miles, by road, south of Blairsden.

This property includes the Gard and Orr, patented claims. No work has been done for a number of years.

Burnham Consolidated Placer Mine. Owner, W. T. Burnham. Location: Poker Flat Mining District; Secs. 5 and 8, T. 21 N., R. 11 E., 4 miles east of Poker Flat, thence 33 miles northerly, by road, to Quincy.

The Burnham Patent contains 60 acres.

Cold Springs Group. Owner, W. L. Wilson, Downieville. Location: Downieville Mining District, in Sec. 12, T. 19 N., R. 10 E., 44 miles by road northwest of Nevada City.

The property consists of the Cold Spring and First Chance claims.

Deer Group. Owners, E. A. and J. O. Hayes. Location: Sierra City Mining District, Sec. 31, T. 21 N., R. 12 E., 7 miles, by road and trail, northwest of Sierra City; 14 miles, by road and trail, south of Blairsden.

The Deer Patent contains 160 acres.

Diana Mining Company. Owner, same. Merchants Exchange Building, San Francisco. Location: On Gibson Creek between Gibsonville and Whiskey Diggings. Gibsonville is 26 miles, by road, south of Quincy.

Dragon Fly Mine. Owners, N. H. Squire, C. J. Simmons, Downieville. Location: Downieville Mining District, Sec. 33, T. 21 N., R. 11 E., 8 miles, by trail, northeast of Downieville, 18 miles to Blairsden by trail and road. Elevation 5000'.

This property consists of three claims, the Dragon Fly, Dragon Fly Extension, and the Pinchbug, containing 60 acres and covering a length along the channel of 1300'. It is situated on the Middle Fork of the North Fork Yuba River, and is a back channel of the present stream covered by a slide. The deposit is about 23' in depth and consists of black wash and some quartz on a slate bedrock. Pay gravel averages about 16' in width and 3' in depth, plus 1' of bedrock. There is a 500' drain tunnel and 300' of drifts, all in gravel. The property was located in 1905 by the present owners, and two men have worked it with a total output of about $3000 to June, 1914.

There is some timber on the property, and water is obtainable under a 120' head, minimum 20", maximum 100". Hoisting is done by water power, and there is electric equipment for lights and blowers.

Adjoining mines are the Wideawake and the Sierra Poorman.

Golden Era Mine. Owner, Richard Moore. Location: Port Wine Mining District, Sec. 18, T 21 N., R. 10 E., 2 miles southeast of St. Louis, thence 33 miles northerly, by road, to Quincy. The Golden Era consists of 103.3 acres of patented ground.

Golden Scepter Group. Owner, G. A. Maxwell, Continental Building, San Francisco. Location: Poker Flat Mining District, near Table Rock, Sec. 22, T. 21 N., R. 10 E., 30 miles south by road, from Quincy.

The holdings consist of the Golden Scepter and Golden Anchor placer claims, and the True Friendship and Golden Sheath quartz claims.

Halsey Group. Owner, W. D. O'Brien, 718 Monadnock Building, San Francisco. Bonded to Wm. H. Loftus, Los Angeles. Location: Sierra City Mining District. Sierra City is 56 miles, by road, northeast of Nevada City, 20 miles, by road, south of Blairsden.

Holdings include the Claybank or Plumas Development Company's property and also the United Consolidated and Excelsior claims.

Hayes Claim. Owner, Ah Gook, Table Rock. Location: Gibsonville Mining District, Sec. 31, T. 22 N., R. 10 E., 1 mile northwest of Table Rock, and 30 miles by road south of Quincy. Elevation 5000'.

Hobby Mine. Owner, M. W, Hobby, Camptonville. Location: Indian Hill Mining District, Sec. 17, T. 19 N., R. 9 E., in lower end of Indian Valley. Indian Hill is about 40 miles, by road, north of Nevada City.

This property is a one-man mine.

I.X.L. and South I.X.L. Group. Owner, Thomas Hayes, Table Rock. Location: Gibsonville Mining District, Sec. 33, T. 22 N., R. 10 E., near Fillmore, about 2 miles northeast of Table Rock and 32 miles south, by road, from Quincy.

Michigan Claim. Owner, L. L. Clough, Quincy. Location: Gibsonville Mining District. Gibsonville is 26 miles south, by road, from Quincy.

Middle Yuba Mine. Owners, Poorman and Hastings, 1724 Santa Clara avenue, Alameda. Location: Pike City Mining District, Secs. 15, 16, 20, 21, and 22, T. 18 N., R. 9 E., 2 miles southeast of Pike City, thence 28 miles southerly, by road, to Nevada City.

The Middle Yuba property consists of 90.35 acres of patented placer ground along the Middle Fork of the Yuba River.

Mount Rainier Claim. Owner, R. J. McCullough, Alleghany. Location: Forest Mining District, at Topohachkin Flat; Secs. 8 and 9, T. 18 N., R. 10 E. Forest is 39 miles by road northeast of Nevada City.

Omega Gold Mining Company. Owners, same, Forest City. Location: Forest Mining District, Sec. 29, T. 19 N., R. 10 E., 2 miles west of Forest City, thence 39 miles southwesterly by road to Nevada City.

This property consists of two claims, the Omega and the Manzanita, the latter having formerly been known as the Lucky Dog.

Pennsylvania Claim. Owner, John Doherty. Bonded to Hong Flat Company. Location: Poker Flat Mining District, Sec. 32, T. 22 N., R. 10 E., 1 mile north of Table Rock, 29 miles south by road from Quincy.

This property consists of one patented claim, the Pennsylvania, with an area of 11.26 acres.

Pilot Peak Mining Company. Owner, Frank Delahunty. Location: Gibsonville Mining District, near Whiskey Diggings, 25 miles south by road from Quincy.

This property is adjoined on the west by the Swift Shore mine.

Poor Boy Claim. Owner, W. W. Casserly, Goodyear Bar. Location: Downieville Mining District, Sec. 32, T. 20 N., R. 10 E., 3 miles west of Downieville, 42 miles by road northeast of Nevada City.

Slate Creek Claim. Owner, C. A. Reed, Bangor, Butte County, California. Location: Gibsonville Mining District, Sec. 36, T. 22 N., R. 9 E., Sec. 31, T. 22 N., R. 10 E., and Secs 1 and 11, T. 21 N., R. 9 E., on Slate Creek at mouth of Gibsonville Creek; 26 miles south by road from Quincy.

This claim extends up the bed of the creek a distance of 1200'.

Slate Creek Debris Dam Company. Owner, J. H. Thomas, Oakland. Location: Poverty Hill Mining District, Sec. 31, T. 21 N., R. 9 E., 2 miles northwest of Mount Pleasant, thence 41 miles northerly, by road, to Quincy.

This property consists of tailings claims on Slate Creek, extending from the Yuba County line to 1600' above it.

Soldier Boy Claim. Owner, M. M. Marshall, 655 Congress street, Portland, Maine. Location: Poker Flat Mining District, Sec. 15, T. 21 N., R. 10 E., 2 miles south of Poker Flat, thence 33 miles northerly by road to Quincy. Elevation 6200'.

Sunnyside Claim. Owner, Franklin Consolidated Placer Gold Mining Company; B. L. Jones, agent, Quincy. Location: Poker Flat Mining District, Secs. 5 and 8, T. 21 N., R. 11 E., 4 miles east of Poker Flat, thence 33 miles northerly, by road, to Quincy.

The Sunnyside placer claim covers 160 acres and is patented.

Tailings Claim. Owners, John Rosenfelds and Sons, 1024 Merchants Exchange Building, San Francisco. Location: Port Wine Mining District, Secs. 25 and 36, T. 21 N., R. 9 E., 2 miles southeast of Port Wine, thence 37 miles northerly, by road, to Quincy.

This claim is located in Cañon Creek, between Craigs and Morristown crossing.

Tough and Hardy Claim. Owner, W. B. Meek, Camptonville. Location: Pike City Mining District, Sec. 6, T. 18 N., R. 9 E., 1-1/2 miles east of Camptonville, thence 53 miles southwesterly by road to Marysville; 1-1/2 miles northwest of Pike City, thence 28 miles southerly, by road, to Nevada City.

Viola Group. Owner, Forest City Mining Company, Downieville. Location: Forest City Mining District, Sec. 19, T. 19 N., R. 11 E., located 3000' east of Cornish Ranch; Forest City is 39 miles northeast, by road, from Nevada City.

This property consists of two claims, the Viola and Viola Extension. [10]

- Depot Hill hydraulic mine has been operated practically every season for many years. The mine is 5 miles north of Camptonville on the state highway running to Downieville in sec. 19, T. 19 N., R. 9 E.

Indian Hill mine near Camptonville was operated during 1931 and 1932. The material consisted mainly of slides from the faces of the old workings. The gravel deposit was 35 feet thick; the grade of bedrock was 1 inch to the foot.

Loftus Blue Lead Mining Company, South Pasadena, operated a hydraulic mine in 1940 and 1941 on a large group of claims running from St. Louis to Howland Flat, a distance of 4 miles by road, in secs. 31, 32, T. 22 N., R. 10 E.; secs. 5, 6, 7, T. 21 N., R. 10 E.; and sec. 12, T. 21 N., R. 9 E. In 1940 the yield from 60,000 cubic yards of gravel was 352 ounces of gold and 25 ounces of silver.

Pioneer Project mine in secs. 13, 14, 23, 24, T. 21 N., R. 9 E., is a consolidation of the old Pioneer with the adjoining Comet, Challenge, and Riffle claims. It was worked in 1942 and 1943. Hydraulic operations for a period of 10 days in 1942 yielded 71 ounces of gold and 4 ounces of silver from 17,000 cubic yards of gravel.

Poverty Hill Properties, San Francisco conducted operations in 1940-41 on a part of the main La Porte channel, one of the old Eocene auriferous channels. The property consisted of 1100 acres in sec. 32, T. 21 N., R. 9 E, and sec. 5, T. 20 N., R. 9 E. The property was first worked by hydraulicking, but later a connected bucket dredge was installed in a pond in one of the hydraulic pits.

William Richter & Sons, Oroville, operated a dragline dredge on the Yuba River in secs. 16, 17, T. 19 N., R. 9 E., in 1941 and 1942. In 1941 operations from June 1 to December 31 yielded 1403 ounces of gold and 179 ounces of silver from 280,000 cubic yards of gravel. In 1942 operations from January 1 to April 30 yielded 243 ounces of gold and 29 ounces of silver from 58,000 cubic yards of gravel.

Ruby mine. There are 1,150 acres, of which 300 are patented, in secs. 10, 11, 14, 15, T. 19 N., R. 10 E. The mine is reached by 15 miles of road from Downieville, mostly steep mountain road, dirt surface.

The Bald Mountain extension channel, one of the oldest Tertiary channels, branches from the Bald Mountain channel at a point north of Forest. Bald Mountain channel is the same as the Ruby and City-of-Six channels. The Bald Mountain Extension channel was worked in the Ruby mine in the nineties from an adit level driven from the side of the mountain on which the town of Forest is located. Present work is on the opposite side of the mountain. A point in the adit is 1,850 feet south of the common corner of secs. 10, 11, 14, 15, T. 19 N., R. 10 E. [1]

STANISLAUS COUNTY

• The San Joaquin River runs across the center of Stanislaus County, and has the Tuolumne and Stanislaus rivers, which flow west from the Sierras, as tributaries. Water for irrigation on the east side of the San Joaquin River is supplied from the Tuolumne and Stanislaus Rivers, the water being diverted by large dams. [2]

• There are no gold mines in the county that are being operated at the present time. There is a gold dredge owned by the La Grange Gold Dredging Company that is operating on the Tuolumne River near La Grange. [3]

Knight's Ferry

Location. The Knight's Ferry district is in northeastern Stanislaus County and western Calaveras County. The town is located on the lower Stanislaus River about 12 miles east-northeast of Oakdale.

History. The district was placer-mined during the gold rush. The town, which was named for William Knight, was an important staging and supply center for the mines and camps of the southern Mother Lode region. It was the seat of Stanislaus County government from 1862 to 1872. The town also was once known as Dentville for the Dent brothers who were brothers-in-law of President U. S. Grant. The old wooden covered bridge that is still standing was reportedly designed by Grant in 1854. From the 1870s through the 1890s, numerous Chinese placer miners reworked the old tailings and small deposits overlooked by the Forty-Niners. There was dragline dredging in the district in the 1930s and early 1940s.

Geology. The east portion of the district is underlain by greenstone and quartz porphyry and the western portion by andesite. The gold was recovered from isolated patches of quartz-rich gravel of Eocene age, younger channel and terrace deposits buried under or adjacent to the Plio-Pleistocene latite of Tuolumne Table Mountain, and Recent gravels in and along the present channel of the Stanislaus River.

La Grange

Location. This district is in southeastern Stanislaus County. It is primarily a dredging field that extends westward from the town of La Grange along the Tuolumne River for nine miles. This district also includes a dredging field two miles to the south, on an older river channel, and surface "diggings" to the north. The town, originally known as French Bar, was founded in 1852, but the name was changed to La Grange in 1856 in honor of Lafayette's country home. For a time Bret Harte taught school here. The estimated output from dredging is $13 million worth of gold.

Geology. The stream gravels in and adjacent to the present Tuolumne River are medium to coarse, loosely consolidated, and average 30 to 35 feet in thickness. The gravels are underlain by tuff. Dredge recoveries during the 1920s averaged 11.6¢ per yard, but later recoveries are believed to have been less.

Minor amounts of platinum were recovered. The dredged area is about nine miles long and 1/2 mile wide. A Pleistocene river channel two miles to the south was dredged for a distance of 1-1/2 miles and is 1/4 mile wide. Here the gravels were compact with a thick volcanic overburden. Thin Eocene gravel patches with abundant quartz boulders and interbedded clay and sand are exposed to the north.

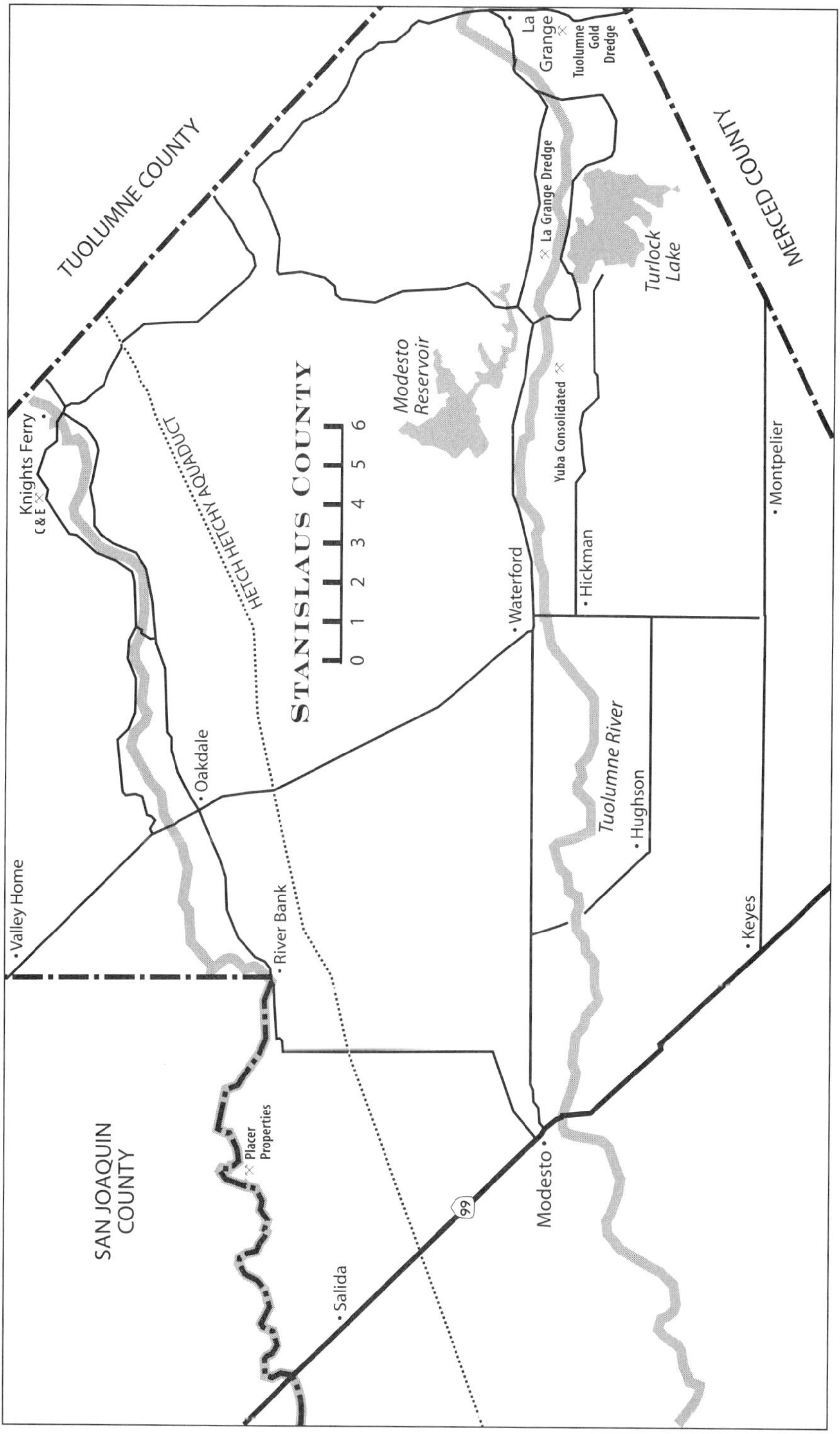
TUOLUMNE COUNTY
MERCED COUNTY
SAN JOAQUIN COUNTY
STANISLAUS COUNTY
0 1 2 3 4 5 6
La Grange
Tuolumne Gold Dredge
La Grange Dredge
Turlock Lake
Modesto Reservoir
Yuba Consolidated
Knights Ferry
C & E
HETCH HETCHY AQUADUCT
Waterford
Hickman
Montpelier
Oakdale
Tuolumne River
Hughson
Valley Home
River Bank
Keyes
Placer Properties
Modesto
99
Salida

Concerns. La Grange Gold Dredging Co., 1907-42 -and 1945-51, one dredge, the longest in the state; Tuolumne Gold Dredging Corp., 1938-43 and 1945-?, one dredge; Yuba Cons. Goldfields, 1941-42, one dredge.

- C & E Dredging Company, Portland, Oregon, operated a dragline dredge on Littlejohn Creek two miles northwest of Knights Ferry intermittently between September 20 and December 17, 1940. [1]
- The La Grange Gold Dredging Company is the only dredging company operating in Stanislaus County.

  This company began operations in 1908 and has one dredge at work. The president of the company is Jesse W. Lilienthal, San Francisco, Cal.; E. A. Wiltsee, New York, is general manager, and Archie Scott is resident manager at La Grange, Cal.

  The holdings of the company comprise an area of about 200 acres of dredging ground, located along the Stanislaus River about four miles below the town of La Grange. The ground is said to average in depth about 35 feet. The gravel is medium coarse, and is overlain by a clay soil ranging in depth from 8 to 10 feet. The bedrock is soft, like at Oroville, and is easily dug by the dredge.

  The ground was prospected by means of drills, and is said to carry more gold than the average gravel at Oroville. The surface contour of the ground is flat and free from timber growth. [2]
- La Grange Gold Dredging Company. Owns one dredge on the Tuolumne River, having 7 cubic feet close connected buckets, shaking screens, Holmes gold-saving tables, belt tailing conveyor and digs 32 feet below water level. It works on spuds. The company owns 200 acres of dredging ground that averages 35 feet deep. The gravel is medium coarse and the bedrock is soft. The ground was prospected by drills. [3]
- Placer Properties, Oakdale, operated a dragline dredge on the Stanislaus River nine miles east of Oakdale throughout 1940. In 1941 operations were continued at a point eight miles east of Oakdale. [1]

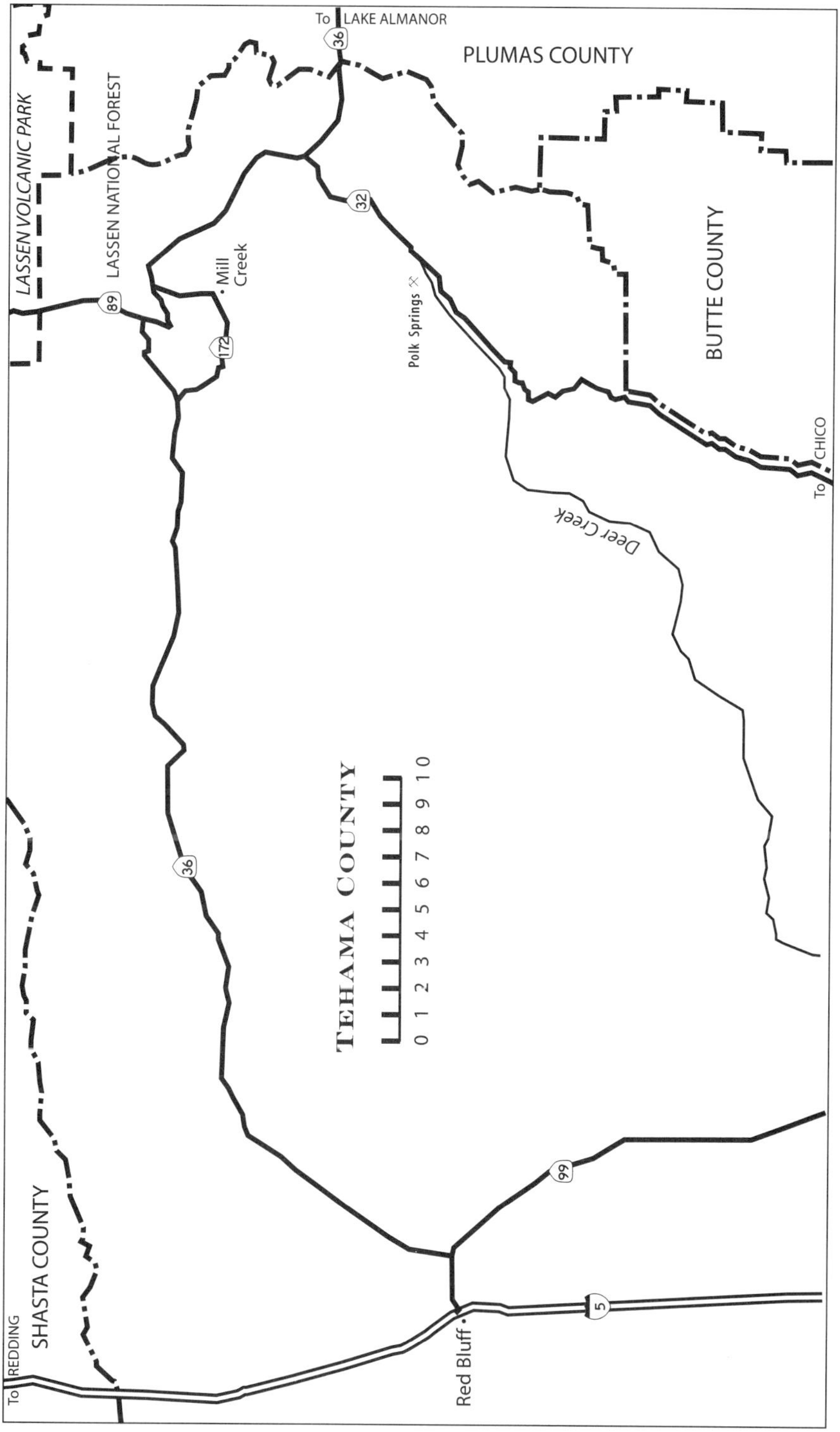
To LAKE ALMANOR
PLUMAS COUNTY
LASSEN VOLCANIC PARK
LASSEN NATIONAL FOREST
Mill Creek
Polk Springs
BUTTE COUNTY
To CHICO
Deer Creek
TEHAMA COUNTY
0 1 2 3 4 5 6 7 8 9 10
SHASTA COUNTY
To REDDING
Red Bluff
36
32
89
172
99
5

TEHAMA COUNTY

Polk Springs

This is a small placer-mining district in eastern Tehama County just south of Polk Springs. It is about 30 miles east-southeast of Red Bluff. Some gold was recovered by hydraulicking years ago, the last work apparently having been done in the 1930s. The deposit consists of gravels up to 50 feet in thickness that contain pebbles of quartz, granitic rocks, and volcanic rocks. Bedrock is schist, which occurs with granitic rocks in a window completely surrounded by volcanic rocks.

TUOLUMNE COUNTY

• GENERAL REVIEW

The placers of Tuolumne County were the richest in California and probably nowhere in the entire world has so much placer gold been derived from an equal area. The following table, showing the value of gold shipped from the principal placer localities of the county by express companies prior to 1899, is but a partial statement of the yield, as other unrecorded millions were taken from the placers.

This table is abstracted from a book prepared for the San Francisco meeting of the American Institute of Mining and Metallurgical Engineers in 1899, at which time records of gold shipments by Wells Fargo & Co., later destroyed in the San Francisco fire, were still available.

Express shipments of gold from principal placer localities in Tuolumne County

Columbia and Springfield	$55,000,000
Groveland, Deer Flat, and Big Oak Flat	25,000,000
Sonora	11,000,000
Don Pedro Bar, Jacksonville, and Stevens Bar	9,000,000
Gold Springs and Nigger Gulch	7,500,000
Shaws Flat	6,000,000
Campo Seco	5, 500, 000
Browns Flat	4,500,000
Poverty Hill and Chili Camp	4,000,000
Pine Log, Experimental Gulch, Italian Bar, Yankee Hill, Knickerbocker Flat	3,500,000
Jamestown	3,500,000
Sullivan Creek	3,000,000
Kincaid Flat	3,000,000
Chinese Camp	2,500,000
Sawmill Flat	2,500,000
Algerine	2,500,000
Mormon Creek	2,225,000
Montezuma and Picayune Gulch	1,500,000
Table Mountain and Humbug	500,000
Horseshoe Bend	300,000
Moccasin Creek	150,000
Total allocated output	**151,175,000**

California Miners' Association, California Mines and Minerals: California meeting, Am. Inst. Min. Eng., San Francisco, 1899, p. 358.

Other prominent producing localities were Six-Mile Bar, Willow Bar, Green Spring Run, Byrnes Ferry, Central Ferry, Peoria Flat, Morehouse Bar, Confidence and Davis Flats, Blanket Creek, Turnback Creek, Cherokee, and Rough & Ready.

All of the allocated output of $151,175,000 listed above was made prior to 1899. Taking into account the additional production of the camps for which the output is not specified and also the known placer production of the county from 1903 to 1937, which amounted alone to about $800,000, the total placer-gold production of the county might reasonably be estimated as approximately $155,000,000.

In the early days the streams and gulches in and adjacent to the Mother Lode district were very rich in gold, but the gravels in the limestone area at Columbia and Springfield area surpassed them all and carried gold in fabulous quantities. Here, in a flat basin less than 2 miles in diameter, over $55,000,000 was recovered between 1853 and 1870 largely by hand methods in shallow gravels that filled potholes and crevices in the eroded limestone bedrock. The placer deposits of the Columbia Basin and the method of their formation will be discussed later. It suffices here to state that probably nowhere in the world, not excepting the famous Eldorado and Bonanza Creeks of the Klondike or the Third Beach Line at Nome, were the deposits so rich.

It may be of interest to draw a few comparisons between the yield of the placers of Tuolumne County and those of famous placer districts elsewhere. According to the Federal Geological Survey, the whole Seward Peninsula of Alaska, embracing the Nome district and eight other principal placer regions with a total area approximately as large as all the placer districts of California combined produced $104,825,000 in gold from the first output in 1897 to 1936, inclusive. This is less than 70 percent of the production of Tuolumne County alone. The rich Fairbanks-Richardson district, outranking all other individual districts of Alaska in its output of placer gold and closely approximating the placer districts of Tuolumne County in size, has yielded $105,741,500 from the time of its discovery in 1903 to 1936, inclusive. This, again, is far surpassed by the production of Tuolumne County.

The gold in the placers of Tuolumne County was very coarse, and it is unlikely that any other area of comparable size yielded so many large nuggets. Nuggets up to a pound in weight were very common, and even today some of fair size are found occasionally.

(Compiled from California Miners' Association, Mining in Tuolumne County; California Mines and Minerals: Am. Inst. Min. Eng., California meeting, San Francisco, 1899, p. 357.)

Over 95 percent of the huge yield of placer gold from Tuolumne County was obtained from quaternary gravels, the gold content of which was derived in considerable part from erosion and reconcentration of gravels of the earlier Tertiary streams. However, the total production of gold from original Tertiary gravels within the county probably has not exceeded 5 or 6 million dollars, derived chiefly from surface placers at Chinese Camp and Montezuma and from drift mines on the Table Mountain channel.

Today the placer deposits of the county rapidly are approaching exhaustion. Although thousands of men were engaged in mining the richer gravel by hand methods from 1853 to 1870, the depletion of the richer deposits led to a gradual desertion of the placers, until in the late 70's the output of placer gold had dwindled to a few hundred ounces a year. Accurate statistics of the placer-gold production before 1903 are not available, but as far as known it was not until 1932 that the annual placer output of the county again became as great as 1,000 ounces. Since then, however, under the stimulus of the higher price of gold, the output has risen steadily, until in 1937 it was 5,720 ounces valued at $200,235.

Virtually all of this recent production has been derived from power-shovel or dragline operations on gravel too low in grade to have been worked profitably at the old price of gold. In the summer of 1938 five draglines with floating washing plants were operating in the county. Of these, two were working on Woods Creek above-the Eagle-Shawmut mine, one on Woods Creek at Jamestown, one on Moccasin Creek, and one at the Mencke-Hess placer near Chinese Camp. In the early fall a small experimental plant in the nature of a floating power

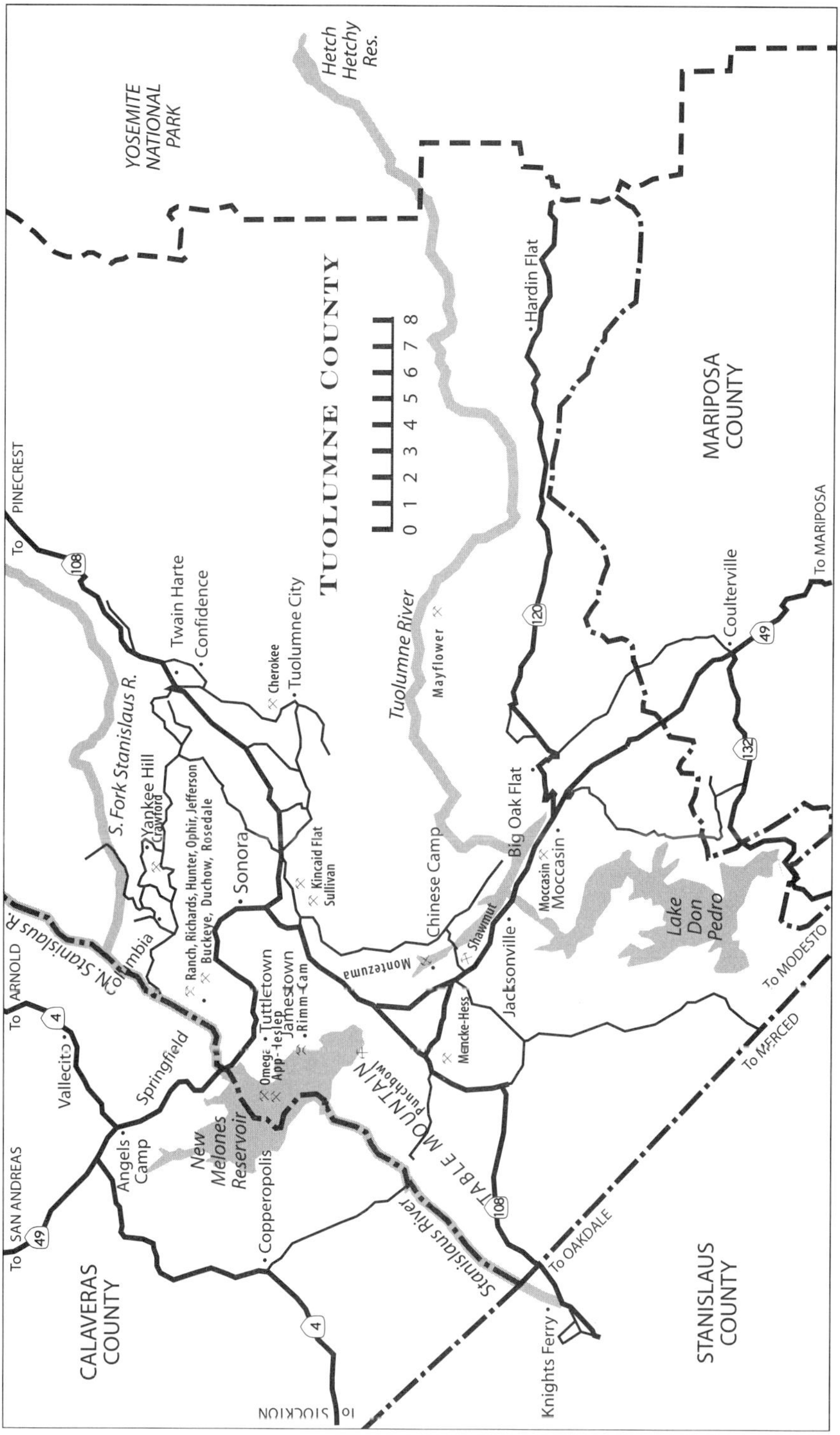
TUOLUMNE COUNTY
0 1 2 3 4 5 6 7 8
YOSEMITE NATIONAL PARK
Hetch Hetchy Res.
MARIPOSA COUNTY
CALAVERAS COUNTY
STANISLAUS COUNTY
To PINECREST
108
Twain Harte
Confidence
Tuolumne City
Cherokee
S. Fork Stanislaus R.
Yankee Hill
Crawford
Ranch, Richards, Hunter, Ophir, Jefferson
Buckeye, Duchow, Rosedale
Sonora
Columbia
N. Stanislaus R.
To ARNOLD
4
Vallecito
Springfield
Angels Camp
To SAN ANDREAS
49
New Melones Reservoir
Copperopolis
To STOCKTON
Tuttletown
Jamestown
Rimm-Cam
Omega
App-Jeslep
Punchbowl
TABLE MOUNTAIN
Stanislaus River
Kincaid Flat
Sullivan
Tuolumne River
Mayflower
Chinese Camp
Montezuma
Shawmut
Mencke-Hess
Jacksonville
Big Oak Flat
120
Hardin Flat
Moccasin
Lake Don Pedro
Coulterville
132
To MARIPOSA
To MODESTO
To MERCED
To OAKDALE
Knights Ferry

shovel was installed on the Tuolumne River near Jacksonville. There are few gravel areas remaining within the county, however, that are suitable for dragline operations and their exhaustion apparently is a matter of only a few years.

Areas suitable for dredging are likewise lacking, though Quaternary gravels are being dredged successfully just beyond the western boundary of the county at La Grange in the alluvial fan of the Tuolumne River where it issues from the mountains. Though some small areas of rich gravel that were overlooked probably will yet be discovered within the county, and other small areas which formerly were unworkable because of water conditions may perhaps now be worked at a profit, the remaining placer-gold resources of the county are very limited and their virtual exhaustion within a few years may be expected.

"Sniping" for gold along the creeks and rivers of the county was done by several hundred itinerant prospectors during the year, and a few of the more fortunate or more skillful among them made fair wages, although in general the average recovery of gold probably amounted to less than 75 cents per day per person. A small gasoline pump was utilized to keep seepage water from the creek out of the pit, and the two partners reported that they recovered over $400 in gold in less than two weeks from a small bank of gravel that had been overlooked by early miners. Such incidents , however, are infrequent.

COLUMBIA BASIN

Nowhere in the world has so much gold been taken from so small an area of placer ground as in the Columbia Basin of Tuolumne County. There, on an open flat, within a radius of a single mile, $55,000,000 in nuggets and gold dust was panned, rocked, or sluiced by miners from 1853 to 1870 and shipped by the Wells Fargo and local express companies. At the present price of gold, recorded shipments from this basin thus represent more than 93 millions of dollars, but the gold unaccounted for in the records of express shipments, which was produced before express service began or was carried away in the pokes of miners, probably would raise this total to 100 millions.

Today, however, there is no active mining within the basin, where only the white pinnacles of the deeply eroded limestone bedrock serve as monuments to the once thriving placer industry and to the eager horde of anonymous adventurers who hopefully toiled on the bedrock between them. The town of Columbia, once proposed as the site for the State capital, an honor said to have been lost by a single vote, is today a somnolent village claiming interest only by its few relics of its romantic past. The smaller placer camps at Gold Spring and Springfield on the circumference of the basin have virtually vanished.

This richest of the world's known placers resulted from a fortunate combination of geologic causes. Natural riffles of irregularly eroded white, crystalline, limestone bedrock about 11/2 miles wide lay in the path of a stream that carried the debris of gold-bearing rocks. The source of the, alluvial debris was nearby in slates such as are seen on both sides of the limestone. These slates contained and still exhibit numerous narrow seams and stringers of gold-bearing quartz with occasional small pockets of exceedingly rich ore. Erosion of the slates, then standing at a considerably higher elevation than at present, by tributaries of the main stream finally brought down to it the gold and quartz together with the softer slates of the hills, which disintegrated more readily. Much of the gold was retained on the limestone riffles of the Columbia Basin, and the larger fragments of quartz sank into pot holes, to mill about until worn away by attrition.

It is possible, of course, that besides the quartz stringers and occasional rich pockets such as are still found in the slates, some large, rich ore shoot may have been eroded away, contributing some of the many large nuggets found in the placers of Columbia Basin; yet in a region of such rich pockets as have been found in the vicinity of the Columbia Channel they alone could account for the coarse gold and large nuggets so abundant in its placers.

Incidentally, it may be mentioned that many bones of the mastodon and of early forms of the elephant and horse were found by the miners in the gravels of the basin, as well as stone and bone implements of some prehistoric race.

The erosion tributary to the Columbia Basin occurred in two periods. The first was in the prevolcanic Tertiary time, when, according to Lindgren, "The drainage was evidently northward by Gold Spring toward some point of the deep channel of the Tertiary Calaveras River near Douglas Flat."

This early stream bed, known as the Columbia Channel, has been traced from a source with a present elevation of 2,060 feet at Yankee Hill, south along Woods Creek, then west across the low divide southwest of Columbia to Mormon Creek, and thence north to Gold Spring, where its elevation is 2,150 feet. Beyond there the channel has been cut away by the canyon of the Stanislaus River, but on the opposite side of the river, near the junction of the Moaning Cave road with the Parrott Ferry-Vallecito highway, at an elevation of about 2,000 feet, there is a remnant of an ancient channel exposed by erosion and filled with gravel composed largely of quartz. It was mined in the past with a reported yield of nearly a million dollars. The altitude and direction of this channel remnant correlate it almost exactly with the Columbia Channel, supporting Lindgren's conclusion that in early Tertiary time the drainage of the Columbia Basin was to the north through a tributary of the Tertiary Calaveras River.

During the andesitic epoch of the late Tertiary the Columbia Basin, in common with the rest of the countryside, was deeply blanketed by tuffs. When a new drainage of the region was developing, the stream that flowed through Table Mountain Channel stole the drainage of the Columbia Basin.

Erosion of the andesitic tuffs during Pliocene and Quaternary time removed most of them. As soon as bedrock was reached, erosion of the gold-bearing veins and concentration of their gold were renewed; but as weathering progressed and the Table Mountain Channel, with its protective capping of lava, was converted into a divide, the drainage of the Columbia Basin was diverted again and today flows into the Stanislaus River through Mormon Creek. Except in minor features, however, the present land surface of the Columbia Basin probably is about as it was in Eocene time.

As the gravels of the basin have been formed locally, they consist largely of subangular and little-worn fragments of quartz. These gravels with their accompanying gold found ideal lodgment in the crevices and potholes of the eroded limestone bedrock and there remained until discovered in 1853.

The early miners excavated the gravel by hand from between the limestone pinnacles of the bedrock and hoisted it in buckets by derricks to raised platforms at the head of sluice boxes. Thousands of men worked in the diggings, and the yield per cubic yard of clay and gravel they handled was enormous. The gold was very coarse, and the finding of a nugget weighing a pound or so was so usual as to attract little attention. However, the early miners were handicapped by lack of drainage for their workings in the flat basin, and as they were unable to mine much below the water level there are probably many deep pockets and holes in the limestone that have never been bottomed.

One such hole, to the east of Columbia, was cleaned out to a depth of over 200 feet, only to disclose an underground torrent at its bottom. Large springs thought to be outlets of subterranean streams flowing through fissures in the limestone gave names to the early placer camps at Gold Spring and Springfield on the rim of the basin.

TABLE MOUNTAIN CHANNEL

Drift mining of gravels from the buried channels of ancient rivers, as described in Bulletin 413, has been much more important in Calaveras County than in Tuolumne. There are only two such channels in the latter—the Tertiary Tuolumne River and Table Mountain Channel.

Table Mountain Channel, also called, as to its upper reaches, the Cataract Channel, was part of a new drainage system developed in the debris from andesitic eruptions of the early Neocene period that had filled the valleys of previously existing streams. It begins near the headwaters of the Middle Fork of the Stanislaus River. Cutting through the easily eroded tuffs and breccias, it took a general southwesterly course, crossing the westerly course of the earlier Tertiary drainage.

Rising in Tuolumne County, Table Mountain Channel crosses the North Fork of the Stanislaus River into Calaveras County 2 miles southeast of Big Trees. There it swings to the southwest and continues down the west bank of the Stanislaus River to Parrott Ferry, intersecting on its way the older Tertiary Calaveras River at nearly a right angle a few miles east of Vallecito. Recrossing the canyon of the Stanislaus into Tuolumne County at Parrott Ferry, the channel runs southeasterly toward Sonora for about 3 miles and then swings almost at a right angle to the southwest to cross the Stanislaus again about 5 miles above Knight's Ferry. Thence it follows the west bank for 2 or 3 miles and again recrosses the river into Tuolumne County; but its course cannot be followed farther downstream, though it probably emptied into the Ione Sea a short distance below Knight's Ferry. Its course in the lower section, where it cannot be followed, is below the level of the present river.

This channel provided drainage for a long time as, in its lower reaches particularly, it not only cut through the earlier volcanic rocks but in some places eroded the underlying Bedrock series to depths of 200 to 300 feet. For example, a, few miles above Knights Ferry the channel bottom is level with the bed of the Stanislaus River, the canyon of which here cuts quite deeply into granitic rocks and greenstones. In the course of its formation, the channel accumulated shallow deposits of gravel that, downstream from where it crosses the Tertiary Calaveras River, contains enough gold to have invited many mining ventures, although in only a few instances were gravels found that were rich enough to pay. Some of the gold in this channel undoubtedly was derived from veins of the Mother Lode and vicinity by direct erosion of the stream; but as this erosion was comparatively slight, it is believed that the greater part of the gold was taken from the rich gravels of the Tertiary Calaveras River and its intersected tributaries, which in general had north-south courses. In this connection it is of interest to note that the Table Mountain Channel apparently stole the drainage of the Columbia Basin from the Tertiary Calaveras River by cutting such a tributary, so that some of the gold in the Table Mountain Channel probably was derived from the immensely rich gravels of this basin.

Another place where the Table Mountain Channel intersected the prevolcanic drainage is just north of Jamestown where the adits of the New York and Humbug mines followed a north-south Eocene Channel from opposite sides of Table Mountain until their workings intersected beneath the mountain, with the result that there was an underground battle for possession of the rich gravel followed by prolonged litigation.

The formation of placer deposits by the Table Mountain Channel was stopped abruptly by renewed volcanic eruptions, which filled the valley of the stream with andesitic muds and tuffs and later, probably in the Pliocene epoch, covered them with a flow of lava 50 to 250 feet thick. This lava is a black latite that is very resistant to weathering. It is characterized by abundant porphyritic crystals of labradorite in a fine-grained ground mass that is largely augite, but it contains a little olivene and vesicular obsidian with magnetite and apatite as accessory minerals.

The andesitic cobble and other volcanic debris that formed the sides of the valley down which the lava flowed were much more easily eroded than the lava, and they have been worn away almost completely, leaving the flow as a capping perched high on top of the ancient stream bed. Because it is protected from erosion by this capping, most of this ancient stream bed on its course south of Parrott Ferry is well above the general level of the surrounding country. The lava flow now appears as a gigantic black wall with nearly perpendicular sides several hundred feet high topping a ridge and extending almost continuously for 20 miles, from a few miles north of Jamestown to a short distance above the bridge at Knights Ferry. This immense wall with its nearly flat top, ranging in width from 500 to 1,500 feet, is one of the most striking topographic features of the county. This portion of it, known as Tuolumne Table Mountain, is mentioned frequently in the writings of Bret Harte and Mark Twain.

The top of Table Mountain ranges from 300 to 500 feet above the general level of the country on the southeast side, but on the northwest side the erosion has been greater, and the valley is 500 to 800 feet below the upper edges of the lava cliffs. Upstream from this mountain the continuity of the lava flow has been broken by erosion so that it is now represented only by a series of island-like patches. However, for 8 miles above Parrott Ferry these remnants of the lava stream lie close together and form imposing columnar cliffs 200 to 300 feet high facing the Stanislaus River.

As a rule, the gravel deposits of the Table Mountain Channel are both narrow and shallow, rarely exceeding 100 feet in width or 15 feet in thickness. Further, they have been disturbed by frequent faulting. For example, at the Omega drift mine near where the channel crosses the Mother Lode there is a series of step faults that strike about N. 30° W., one of which has lifted the downstream portion of the channel approximately 60 feet above the upstream section.

As in most stream valleys of considerable age, in some places there are several channels at varying levels between the main rims. In these channels the auriferous gravel normally is covered with 3 to 30 feet of clay, then with 50 to 75 feet of unconsolidated volcanic sand, next by a compact andesitic tuff, and finally by the heavy capping of latite. However, in some instances the layers of clay and sand are missing and the andesitic tuff is in direct contact with the gravel.

The gravel itself is composed of a heterogeneous assortment of rocks not only derived from the primary erosion of the many different rocks cut by the 50-mile channel but containing Eocene gravels robbed from the older drainage. The gold is coarse, but the "pay" areas are extremely spotty. This fact, together with the shallowness of the gravels, their small width, their disturbance by faults, and the normal presence of heavy clay and unconsolidated sands above them, making the mine workings difficult to support, all have contributed to the many failures that have attended attempts to mine this channel.

Most of the mines are confined to a 10 mile stretch under Table Mountain, extending from near Columbia south to Montezuma. Gold is said to have been discovered in the channel in 1854 at a point near its south rim in the vicinity of Shaws Flat, where the lava has been denuded. During the next 10 years more than 40,000 feet of drifts were run to gain access to the channel, and active mining was carried on at 20 or more properties, prominent among which were the App, Boston, Buckeye, Duchow, Eureka, Hunter, Jefferson, Leap Year, Maine Boys, McKinley, Montezuma, Punchbowl, Omega, Ophir, Ranch, Richards, Rosedale, and Springfield. By 1870, however, drift mining here had virtually ceased, although some little work has been done almost every year since. It is estimated that slightly more than $1,000,000 in gold was recovered from Table Mountain Channel. Only one or two of the mines were profitable and it is doubtful if the total recovery equaled the total cost.

When the Table Mountain area was visited in the summer of 1938 work was being carried on at only one drift mine—the Punchbowl—which is described briefly.

PUNCHBOWL

The Punchbowl mine, comprising about 200 acres, is situated in sec. 20, T. 1 N., R. 14 E., and extends completely across Table Mountain about 2-1/2 miles south of the Rawhide mine. It was one of the early drift mines beneath Table Mountain and is being reopened by G. B. Miller, 1825 Highland Place, Berkeley, Calif., and associates.

An adit 1,800 feet long runs east into the west side of Table Mountain to tap the auriferous gravels of the old Neocene streams. About 1,450 feet from its portal a 28-foot raise has tapped an ancient channel that has been drifted upon upstream to the north for about 800 feet. The gravel extracted from the drifts is reported to have averaged $2 per cubic yard and was uncemented. The gold was coarse and well worn, and nuggets weighing 1/4 to 1/2 ounce each are said to have been found frequently. The gravel, consisting largely of well-rounded porphyry, granite, and schist but containing very little quartz, was washed in a plant on the hillside just below the mouth of the adit, but the plant has long since disappeared. It is evident from the small quantity of tailings that very little gravel was breasted in the workings. North of the adit is an old shaft 420 feet deep and sloping about 18° almost due south. A survey has shown that if this shaft were extended 240 feet its bottom would be on the same level and within 100 feet of the north end of the old drift. If a connection were made between the shaft and drift it would provide good ventilation for the workings. At the time the property was visited in the fall of 1938 two men were engaged in removing caved material from near the end of the adit in preparation for making the connection referred to as a means to resume exploratory work.

TERTIARY TUOLUMNE

In Tuolumne County evidences of only one large prevolcanic Tertiary river comparable to the Tertiary Calaveras, Mokelumne, and American Rivers in the Mother Lode counties to the north have been preserved. This ancient stream is known as the Tertiary Tuolumne River. It is the most southern of the Tertiary rivers that drained the Mother Lode region, of which enough evidence remains to allow its general course to be traced with a fair degree of accuracy. Doubtless there were other prevolcanic streams in Tuolumne County and also to the south in Mariposa County that had their sources near the crest of the Sierras and emptied into the Ione Sea. However, the great flows of lava and volcanic mud that filled the valleys of the early Tertiary rivers to the north extended only over the northern portion of Tuolumne County, and in consequence the ancient rivers to the south, lacking this cause of deflection and protective covering against erosion, simply deepened their channels during Quaternary time with little alteration of their courses and have not, therefore, been preserved as distinct streams.

The Tertiary Tuolumne River rose near Mount Dana in the High Sierras in the southeastern corner of Tuolumne County and flowed in a general westerly direction through Hetch Hetchy Valley and on to the north of Groveland and Shawmut through Chinese Camp to empty into the Ione Sea, probably in the vicinity of Knights Ferry. As it was near the southern edge of the area covered by the andesitic eruptions of the late Tertiary in a region blanketed by a relatively thin covering of volcanic ejecta, its valley was not filled completely by the ashes, mud, and lava that flowed into it, and in consequence its course was little altered by the eruptive material. Hence, the later drainage followed the course of the prevolcanic stream closely throughout most of its length. However, during the late Pliocene and early Quaternary, when erosion was speeded by torrential grades resulting from the westerly tilting of the region, the stream was deflected considerably in its lower reaches where it occupied a wide and open valley. There, just downstream from the confluence of the North Fork of the present river and a few miles east of Chinese camp, the old stream was turned from its general westerly course. It cut a new channel to the south, which has since been deepened until, at the confluence of Woods Creek near Jacksonville, the nearest point of the present river to Chinese Camp, it is about 700 feet below the old stream bed at Chinese Camp. Farther upstream the difference in elevation between the present and prevolcanic stream beds is even greater, reaching 1,500 feet or more, showing the great influence that the tilting of the region has had on canyon cutting by the river during Pliocene and Quaternary time.

The geologic history of the Tertiary and present Tuolumne Rivers has an important economic bearing on their placer deposits. Those of the ancient stream consists of patches of gravel remaining here and there between Piute Canyon near the headwaters of the river and Big Humbug Creek, 33 miles farther downstream. These gravels occasionally have been productive, but as in its upper reaches the ancient river traversed granitic rocks that contained few auriferous veins, the Tertiary gravel deposits in this section are relatively unimportant. However, 2 or 3 miles above Big Humbug Creek the old river cut the auriferous veins of the East belt, and 13 miles farther downstream, about 2 miles east of Chinese Camp, it cut the Mother Lode itself and the gold-bearing Mariposa slates that flanked it on both sides.

The gravels resulting from this erosion were unquestionably rich in gold, but they have been almost wholly swept away by more recent erosion, only one considerable area of the ancient gravel remaining at Chinese Camp. Here, within an area of less than 1 square mile, shallow patches of gravel composed largely of quartz, filling the depressions of a rather flat bedrock, are credibly reported to have yielded $2,000,000 to the early miners; and later, after the best of the gravel had been worked, this same area supported several thousand Chinese miners for a number of years, the Chinese having been engaged in cleaning bedrock and working by hand the poorer gravels left by the white miners.

Much of the gravel of the Tertiary Tuolumne River and its tributaries has been eroded and reconcentrated by the present river system and enriched further by Quaternary erosion of the gold-bearing areas. Hence, the placers of the present river within and downstream from the area of gold-bearing rocks were rich and were mined with surprising thoroughness by the old-timers. By building wing dams when the water was low during the late summer and fall, they were even able to work gravels well out in the main river; and the placers of the main stream and such tributaries as Woods, Sullivan, Curtis, Rough & Ready, Blanket, Turnback, Moccasin, Deer, and Big Humbug Creeks and the North Fork were rapidly worked out and yielded millions of dollars.

There were gravel areas, of course, that contained too little gold to be worked profitably by the old-timers, and some of these, in conjunction with tailings from the past workings, now furnish profitable ground for dragline operations. During the summer of 1938 there were four such areas on tributaries of the Tuolumne River, all of which had been mined before, that were being reworked by draglines. Three of these were on Woods Creek and one on Moccasin Creek. Descriptions of the operations follow.

DRAGLINE OPERATIONS

SHAWMUT PLACER—In the summer of 1938 two draglines and floating washing plants were being operated by E. A. Kent, 149 California Street, San Francisco, at the Shawmut placer on Woods Creek about half a mile above the bridge that crosses it on United States Highway 49. Plant 1 was at work in the valley of the creek just above the Eagle-Shawmut mill, and plant 2 was operating about a quarter of a mile farther upstream, both on claims leased from the Woods Creek Placer Co. in sec. 11, T. 1 S., R. 14 E. J. V. Rice of Sonora was general superintendent.

Plant 1 was one of the first draglines and floating washing plants built by the Bodinson Manufacturing Co. and began mining near Oroville, Calif., in January 1936. It was moved to Tuolumne County in October of that year and, after working out about 50 acres of placer at Montezuma Flat, was transferred to the lower end of Woods Creek, where it began operations in June 1937. Here it mined about 600,000 cubic, yards of gravel below the highway bridge and was then completely rebuilt and put in operation again above the bridge in May 1938.

Plant 2 was erected on the Tuolumne River just above the mouth of Woods Creek on November 20, 1937, but was completely demolished 20 days later by a flood of unusual severity, which not only wrecked the washing plant, dragline, and bulldozer but carried away the first clean-up. However, the plant was rebuilt, and mining in the river was resumed on February 20 and continued until April 3, when the plant was moved to Woods Creek, beginning work only 9 days later.

The wash is heavy and well rounded. It contains many cobbles 18 inches to 2 feet in diameter and rests on a slate or schist bedrock. The gravel ranges from 3 to 20 feet in depth, averaging about 10 feet. Two feet of the bedrock is mined where possible.

Woods Creek has been a prolific producer of placer gold, and although the claims under lease have been worked and reworked by hand methods many times the average yield of the ground is large enough to justify the present operation. In June 1938, 600,000 cubic yards of gravel was still available for mining on these claims. Both Plants are operated three 8-hour shifts. Forty men are employed, six of whom are engaged in testing a placer 1-1/2 miles upstream, to which one of the plants may be moved later.

RIMM-CAM DREDGING CO.—In June 1938 the Rimm-Cam Dredging Co., of which Albert and Cedric Rimmer and Jack Cameron, all of Jamestown, Calif., are the active partners, was mining a 6-acre tract of gravel on Woods Creek at Jamestown. Three other small tracts of gravel on Woods Creek between Jamestown and Sonora, totaling 16 acres, have been obtained by the company already, and it is expected that the plant will be busy for several years in mining these and other small areas of workable gravel that are available in the county.

Although the area being mined was worked in the past, considerable coarse gold and occasionally nuggets weighing up to 1/4 ounce are recovered. The gold is 850 fine. The gravel deposit, which is situated largely on the north bank of Woods Creek, ranges from 3 to 6 feet in depth and contains only a few boulders too large to be handled by the stacker. The bedrock is schist. The brush and the few trees growing on the tract are cleared before the gravel is excavated. The stumps are handled by the dragline.

The average yield in gold is reported as 68 cents per cubic yard. Three men are employed, and the plant is operated during an 8-hour day shift only.

MOCCASIN CREEK PLACER—Larsen & Harms, of Sacramento, Calif., are operating a dragline and floating washing plant near the mouth of Moccasin Creek, about 2 miles southeast of Jacksonville, in sec. 20, T. 1 S., R. 15 E. Moccasin Creek has its source near the top of the divide between Tuolumne and Mariposa Counties, where it is crossed by United States Highway 49. The creek is about 7 miles long and flows northwesterly. Near its headwaters it cuts the Mother Lode and then courses along the northeast side of the Lode for 4 miles to its confluence with-the Tuolumne River. The creek has eroded a deep V-shaped valley through the Mother Lode and adjacent rocks, and the gold in its placers was derived from this source. In the past, when its more accessible, gravels were mined and washed by hand it is said to have yielded but 7,500 ounces of gold. The floor of the creek valley for 2 miles above its mouth averages about 350 feet in width and is covered with approximately 10 feet of coarse gravel containing many large cobbles and boulders, the latter often being 3 feet or more in diameter. All this material is well-rounded, and the wash contains little clay, as the grade of the creek is steep enough to keep fine sand and mud in suspension. The bedrock in this portion of the creek is serpentine and slate.

Larsen and Harms began mining about three-fourths mile above the mouth of the creek in November 1937. A strip of ground about 175 feet wide between the center line of the valley and the north bank and extending to the mouth of the creek was first excavated and washed. When the plant was visited in June 1938 mining upstream along the south bank was well under way, the mouth of the creek having been mined completely across.

The gold is 820 to 845 fine and of medium size, with occasional nuggets weighing 1 or 2 pennyweight. A nugget worth $5 is the largest that has been recovered. The plant is operated with three 8-hour shifts of four men each. In June 1938 enough gravel was still available to supply the plant for a year.

MENCKE-HESS PLACER—The Mencke-Hess Placer is about 2 miles south of Chinese Camp in secs. 15 and 22, T. 1 S., R. 14 E., at the head of Sixbit Gulch, which drains into the Tuolumne River 2 miles farther south. Although this valley was formed during Quaternary time, much of the gravel in it probably was derived from the erosion of older placers deposited by the Tertiary Tuolumne River in the vicinity of Chinese Camp, and much of the gold in the present deposit is thought to have come from this source. The gravel is fairly coarse and uncemented and it contains many cobbles weighing from 25 to 100 pounds each; these are not only too well worn to be of local origin, but many are composed of rock found far up the canyon of the Tuolumne. Near the head of the valley the gravel is from 50 to 75 feet deep and is reported to average more than 50 cents a cubic yard. The gold is largely concentrated, however, on and near the bedrock. It is said that in the early days about $40,000 was recovered along the bedrock by drift mining.

In 1935 and 1936 Mencke-Hess Gravels, Inc., under the direction of Lloyd Root, former State mineralogist of California, worked the upper end of the deposit by open-pit methods. It is reported that $92,820 was recovered from 160,000 cubic yards of gravel excavated from the pit, an average of 58 cents a cubic yard; but costs and royalties together exceeded the recovery, and when in the winter of 1937 the banks of the pit caved, mining was abandoned.

The next attempt at mining was about 1/2 mile farther down the valley, where a floating washing plant and dragline were installed Here drilling is reported to have disclosed a body of gravel 250 feet wide and about 30 feet deep, averaging over 50 cents per cubic yard, but again with most of the gold on bedrock.

A cut 85 by 90 feet wide was mined upstream for a few hundred feet, but the serpentine bedrock proved much too hard to be dug, and its surface was so uneven that very little of the rich gravel on it could be scraped into the dragline bucket. In fact, it is reported that only 5 or 6 cents per cubic yard, or about 10 percent of the gold as recovered. After several disappointing clean-ups the pond was pumped down, and an attempt was made to clean bedrock with a suction hose from the intake of a 6-inch sand pump installed on the bank of the dragline pit just above water level. Although some gold was recovered by this method, the immobility of the arrangement foredoomed it to failure and mining again was abandoned in the summer of 1938.

The failure of a well-equipped dragline plant on this ground emphasizes the fact that a diggable bedrock is a primary requisite for a successful dragline operation where the gold is concentrated on or near it.

MISCELLANEOUS PLACER OPERATIONS

KINCAID FLAT AND SULLIVAN MINES—The Kincaid Flat and Sullivan placers are 2 miles southeast of Sonora in sec. 7, T. 1 N., R. 15 E. They are owned by the M. McCormick Co. of Stockton, Calif., and are under lease to James Morrow of Sonora.

Kincaid Flat lies between Sullivan and Curtis Creeks and is the site of an ancient stream whose gravels were rich in gold. The average elevation of the flat is 100 feet or more above the creeks on each side. In the past a tunnel 1,200 feet long was driven from the southeast bank of Sullivan Creek through the limestone rim of this old stream valley and entered the gravels at a depth of about 60 feet and about 90 feet above the lowest point in the bedrock reached by an exploratory shaft. The tunnel was equipped with a sluice 4 feet wide, and after the gravel above the tunnel level was washed, some of the deeper gravels near by were hoisted into the sluice, but much of the deeper ground remains unworked. This early operation is said to have yielded most of the $3,000,000 output with which Kincaid Flat is credited.

Sullivan Creek, a southwesterly tributary of Woods Creek, is reported to have produced an equal amount of gold, but the early miners were unable to work the deep gravels in a bar covering 3 or 4 acres just below Arroyo Seco. These deeper gravels rest on a stratum of limestone, which crosses the creek here and has been eroded far below the level of the harder rocks downstream, which dam the subsurface water so that the gravels in the limestone pocket are flooded to a depth of 40 to 50 feet.

In March 1936 Morrow leased this bar and the Kincaid Flat property and proceeded to test the gravel of the bar by sinking five shafts with the aid of a 1/2-cubic-yard clamshell dipper. In general, pay gravel was reached at a depth of 12 feet, and from this level to a depth of 25 feet the gravel contained $2 to $10 per cubic yard. Bedrock was not reached in any of the shafts, as the available pumping facilities could not handle the water at greater depth. On the strength of this sampling Morrow subleased the bar to Larsen & Harms of Sacramento in November 1937. In January 1938 this company installed a washing plant on the southeast bank of Sullivan Creek capable of handling 30 cubic yards of gravel per hour and began stripping with a 3/4-cubic-yard electric dragline. Unusually heavy winter rains made operations difficult, and continuous work was impossible. Water was pumped from the pit by a 10-inch centrifugal pump mounted on a float and driven by a 50-horsepower electric motor. This pump was unable to lower the water beyond 35 feet, and pay dirt was dug almost continuously through 6 or 8 feet of water. As the underlying limestone has been eroded very irregularly, with the formation of deep crevices and potholes with pinnacles between them and residual boulders of limestone where the pinnacles had been cut through underneath, it was impossible to excavate the gravel with the dragline, particularly as the operator could not see where he was digging beneath the muddy water. Consequently, the sublease was abandoned in May 1938 after $8,000 had been recovered from about 4,000 cubic yards of gravel, which included much overburden dug from the banks of the pit in filling the dragline bucket. The gold recovered was coarse, ranging in size from a pinhead to nuggets weighing half an ounce. It averaged 910 fine.

Morrow plans to work the bar by fluming those sections of the creek that apparently are supplying most of the seepage into the pit, and then pumping out the water with an electric pump and excavating the gravel by hand methods. If this operation is successful, plans will be outlined for working Kincaid Flat. [7]

• GOLD MINES–PLACER.

Buckeye Channel (Rosedale) Mine (drift). It is 3 miles north of Jamestown, on the east slope of Table Mountain. On the Rosedale Channel, having a course northeast and southwest. The bedrock is a soft Calaveras slate. The width of gravel is 60 feet, with a depth of 3 feet. The gravel is worked through incline shaft 325 feet deep. Twelve men are employed drifting on channel. W, F. Wilson, of San Francisco, owner.

Cherokee Gravel Mine (drift). Situated 1-1/2 miles north of Tuolumne, on ridge west of Turnback Creek. Course of channel is northwest and southeast and is capped by andesite flow 400 feet thick. The bedrock is a decomposed granodiorite. Tunnel 400 feet long. Two men are employed retimbering tunnel. Cherokee Gravel and Gold Mining Company, Soulsbyville, owner. Andrew Scanavino, superintendent.

Columbia Basin Mine (drift). Situated 1 mile north of Columbia, in Columbia basin. On the Table Mountain channel. The course of channel is northwest and southeast, the gravel occurring in deep crevices of the limestone formation. The gravels are subangular, little worn and largely quartzose. Shaft 200 feet deep. Three men are employed unwatering mine. Columbia Basin Development Company, 652 Monadnock Building, San Francisco, owner.

Crawford Placer (hydraulic). Situated 1 mile northeast of Columbia, on Yankee Hill. Controls 420 acres on Woods Creek. Course of channel is northeast and southwest, having a depth of gravel of 10 to 80 feet, with a width of 2000 feet. T. C. Crawford, of London, England, owner.

Gold Ship Gravel Mine (drift). Situated 5 miles northeast of Groveland, on ridge north of Columbia Creek. Gravel capped by an andesitic breccia. The gravel is slightly cemented, containing boulders of andesite, quartz, quartzite and slate. Developments consist of an incline shaft 625 feet deep on west rim of channel; on the east rim a shaft was sunk on an incline to a depth of 150 feet. There is a Dodge rotary mill on property. Idle. Gold Ship Mining Company, Ray Building, San Jose, owner. W. J. Graham, manager.

Humbug Gravel Mine (drift). Situated 2 miles north of Jamestown, on the east slope of Table Mountain, on the Table Mountain channel. The channel is capped by basaltic flow 200 feet in thickness. Work is in progress on this property developing the Humbug and Rosedale channels. The Rosedale channel has an elevation of 12 feet above the Humbug Channel and spills into the latter. These, channels are developed through two inclined shafts with depths of 365 and 325 feet. There is a drift 1676 feet long, connecting the incline shafts. The course of channel is northeast and southwest. The width of gravel on Humbug channel is 200 feet, with a depth of 3 to 5 feet. The Rosedale channel has a width of 60 feet, with a depth of gravel of 5 feet. The bedrock is a soft slate. The gravel is cemented and free, composed of quartz, quartzite and andesitic boulders. There is an extensive amount of underground workings on the two channels. Six men are employed. California Company, 60 State Street, Boston, Mass., owner. J. L. Whitney, manager.

Punchbowl Gravel Mine (drift). Situated 6 miles west of Jamestown, on southwest slope of Table Mountain. The course of channel is north and south, and is capped by 250 feet of basalt. Tunnel 1100 feet long being driven in Calaveras slate and diabase bedrock to cut below upper Table Mountain channel. Four men employed. Punchbowl Gravel Mining Company, Paris, France, owner. E. C. Shafer, general manager.

Saratoga Placer Mine (drift). Situated 1-1/2 miles south of Columbia, on the east slope of Table Mountain. Course of channel northwest and southeast, width is 300 feet, with a depth of gravel of three feet. The bedrock is slate. Channel capped by a basalt flow 200 feet thick. Workings consist of an incline shaft 200 feet deep and a drain tunnel 1000 feet long, which intersects shaft at the 200-foot level. Three men are employed. Saratoga Mining and Development Company, 875 Willow Street, Oakland, owner. J. F. Dieu, superintendent.

Stanislaus Gravel (If I Can) Mine (drift). Situated 12 miles southwest of Jamestown, on the Stanislaus River, on Table Mountain channel. Two channels have been developed, the pre-andesite channel (lower) and the andesite channel (upper). Channel capped by basalt flow 135 feet thick. Course of channel is northeast and southwest, width of gravel is about 100 feet with a depth of 8 to 15 feet. The bedrock is greenstone, hard and highly creviced. The gravel is partly cemented and free, containing boulders of quartz and andesite. The gravel is worked through a bedrock tunnel 1500 feet long. From 12 to 40 men are employed. Stanislaus Gravel Mining Company, 1010 Merchants National Bank Building, San Francisco, owner. A. G. Kullberg, manager.

Springfield Placer Mine (drift). Situated on ridge southeast of the Stanislaus River, near Columbia. Seven distinct channels have been developed on this property-Big Buchanan, Ranch, Fox, Clark, Rancheria, Barry, and Slaughter House. The main channel, called the Buchanan, enters the property on northwest corner, passing through the property in a southeast direction for a distance of 11 miles and is over 800 feet wide. These channels have been developed by three incline shafts; on the east slope of Table Mountain an incline shaft has been sunk to a depth of 240 feet; further south a shaft 900 feet deep on 18° incline, and 1100 feet west a shaft cuts the channel at a depth of 240 feet. On ridge above the Stanislaus River a 6000-foot tunnel is being driven under these channels. The formations are granite, slate and limestone. By means of this tunnel the company expects to drain the heavy flow of water encountered in the shafts, which prevented working the rich gravel found there. Thirty men are employed. Springfield Tunnel & Development Co., Springfield, via Columbia, Cal., owner. A. L. Horner, manager; P. Cadematori, superintendent.

American Camp

Location. This district is in northwestern Tuolumne County in the general vicinity of American Camp Station, which is about eight miles northeast of Columbia. It includes the Italian Bar, French Camp, Star Ridge, Grant Ridge, and Cedar Ridge areas. The famous Columbia placer-mining district adjoins it on the southwest.

History. The streams in the district were first mined during the gold rush. The town of Italian Bar was the principal settlement at that time. It later was destroyed by fire. Lode mining began about 1860, and continued almost steadily until around 1900. There was some mining in the district again in the 1920s and 1930s, and there has been minor prospecting since. The Grant mine was prospected for uranium in 1953-54.

Geology. The district is underlain predominantly by argillite, quartzite, siliceous schist, and limestone of the Calaveras Formation (Carboniferous to Permian). There are also a few small granodiorite stocks. Fine-grained diorite and aplite dikes are common and are often associated with the gold-quartz veins.

Ore deposits. Numerous gold-quartz veins and stringers contain small to medium-sized ore shoots. The veins strike either west or north-northeast. In places the ore is rich. The ore contains free gold associated with varying amounts of sulfides, especially galena and chalcopyrite. There are several patches of Eocene channel gravels that have been mined by hydraulicking. At the Grant mine black uraninite associated with gold occurs in quartz.

Mines. Argentum Consolidated, Black Bear, Contention, Gold Ridge, Grant, Gray Eagle, Ham and Bitney $100,000, Hazel Bell, Indian Girl, Keltz $300,000, Lucky Strike, Mountain Lily, Noonday, Rifle, Sonnet, Star, Tiffany, Volunteer.

Big Oak Flat

Location. The Big Oak Flat district is in the East Gold Belt of the Sierra Nevada in southwestern Tuolumne County. It includes the Groveland, Deer Flat, and Second Garrotte areas to the east.

History. This district was first mined shortly after the beginning of the gold rush by James D. Savage. It was named for an oak tree that had a trunk with a diameter of 11 feet. The placer deposits here were highly productive, those at Big Oak and Deer Flat having been credited with a production of $25 million. Lode mining began soon afterward, and continued steadily until World War I. The town of Garrotte was renamed Groveland by later residents who replaced the name given in 1850, when a thief was hanged there. But a place to the east, named Second Garotte (after 1850) for a similar reason, has kept its 19th century name. This area has prospered from tourist trade that originates along the Big Oak Flat road, which serves Yosemite National Park. The Hetch Hetchy Railroad, which was used in the construction of the San Francisco Water Department's Hetch Hetchy Reservoir to the east, served the area for some years. Some lode mining was done in the district again in the 1930s, and there has been intermittent prospecting and development work at several mines since.

Geology. The area is underlain by a northwest-trending belt of schist, argillite, and quartzite of the Calaveras Formation (Carboniferous to Permian), intruded by several small granodiorite stocks and flanked on the west by amphibolite, slate, and serpentine of the Mother Lode belt.

Ore deposits. A considerable number of north-, northwest- and west-trending quartz veins are found in both the metamorphic rocks and the granodiorite. The ore shoots usually are limited in extent, but often are rich. The ore contains free gold and often abundant sulfides, especially galena. Most of the veins are only a few feet thick. The surface placers mined during the gold rush were extremely rich.

Mines. Bicknell, Champion, Contact, Criss Cross, Del Monte, Goodnow, Kanaka, Long Fellow $100,000+, Mack, Mississippi, Mohrman $40,000, Mt. Jefferson, National, Noname, Red Jacket, Rhode Island, Venus, Wide West.

Chinese Camp

Chinese Camp is in western Tuolumne County about 10 miles southwest of Sonora. It was a placer-mining center settled by Chinese miners in 1849. Much work was done in the 1850s, and the piles of soil and gravel turned over in the search for gold can still be seen in nearly every gulch. The old mining town of Chinese Camp is quite well-preserved.

Much of the placer gold was recovered from extremely rich quartzitic gravels of Eocene age. Some gold also was mined from gravels of late Tertiary age. Most of the early-day work was done by hydraulicking and ground sluicing; during the 1930s there was some dragline dredging. Bedrock consists of serpentine, greenstone, and slate. There are a few gold-quartz veins in the area. The value of the total output from placer mining is estimated at $2.5 million.

Columbia

Location. This famous placer-mining district is in north-central Tuolumne County, in the vicinity of the old mining town of Columbia, five miles north of Sonora. It includes the Yankee Hill, Sawmill Flat, Squabbletown, Brown's Flat, and Springfield areas. The Sonora district is just to the south and the American Camp district lies to the northeast.

History. Columbia was one of the richest and most famous placer-mining districts in California. Early in 1850 a group of Mexican miners who had been forced off their claims at Sonora struck it rich here. Americans moved in and in turn forced them to leave. For a short period, the district was known as Hidreth's Diggings and American Camp, but it soon became "Columbia, Gem of the Southern Mines". During the 1850s and early 1960s, the diggings were enormously productive, the output averaging $100,000 or more per week. Columbia was one of the largest cities in California at this time, with an estimated population of 25,000 to 30,000. The district declined in the late 1860s, but small-scale mining continued until recently. The central portion of the old town became a state park in 1945 and is now a popular tourist attraction. Many of the famous old buildings have been restored. The value of the total production of the district has been estimated to be at least $87 million, and some have put the figure as high as $150 million.

Geology. Columbia lies in a preserved Tertiary valley with pre-volcanic features. It is a flat valley that is underlain chiefly by crystalline limestone and dolomite of the Calaveras Formation (Carboniferous to Permian). The limestone has numerous deep potholes and cavities, which contained enormously rich gravel. Several very large nuggets and gold masses were taken here, including one that weighed over 50 pounds and several weighing more than 300 ounces. Slow degradation of the area in pre-volcanic times tended to concentrate coarse gold in this flat basin. It is south of the main Tertiary Stanislaus River. Vertebrate fossils were found in the gravels. In the early day mining operations, the gravels were hoisted from the potholes and washed through sluices and long toms on raised platforms.

Confidence

Location and History. The Confidence district is in west-central Tuolumne County between Twain Harte and Long Barn and about 15 miles northeast of Sonora. It is in the Sierra Nevada East Gold Belt and adjoins the Soulsbyville district on the northeast. This region was first placer-mined during the gold rush, and lode mining began shortly afterward. The Confidence mine was worked on a large scale until around 1912. Some of the mines were active in the 1930s, and the district was later prospected for tungsten.

Geology. The mines in this district are located chiefly in a southwest-extending arm or "peninsula" of granitic rocks of the main Sierra Nevada batholith. The principal rock is granodiorite; slate, mica schist, phyllite, and quartzite lie to the south. Fine-grained dioritic dikes are present and commonly are associated with the gold-quartz veins. Also present are small bodies of tungsten-bearing tactite, which occur as roof pendants in the granitic rocks.

Ore Deposits. The ore bodies occur in north- to northwest-striking quartz veins that usually range from one to five feet in thickness. The ore contains free gold and often abundant sulfides, especially galena, which often is associated with gold. The Confidence vein was mined to an inclined depth of 1200 feet.

Mines. Casa Madera, Corona, Confidence $4.25 million, Excelsior $420,000, Fair Oaks, Gem, Geraldine, Green $200,000, Humbug, Lucky Strike, Morning Glory, Red Cloud, Ripperton, Ryan, Thunderbolt, Too Far North, Wall Street.

Coulterville

See text under Mariposa County, p. 92.

Granite Springs

See text under Mariposa County, p. 93.

Gravel Range

See text under Mariposa County, p. 93.

Hardin Flat

Location. This is a small Sierra Nevada east gold belt district in south-central Tuolumne County. It is west of Yosemite National Park on the Big Oak Flat road and two miles east of the town of Hardin Flat. The town was named for "Little Johnny Hardin", an eccentric Englishman who once owned a sawmill here.

There are a number of small prospects that are intermittently worked, mostly by weekend prospectors.

Geology. The principal rock in the area is granodiorite that is cut by narrow aplitic dikes. There are a number of narrow quartz veins that in places have yielded small but rich pockets of gold near the surface. Some sulfides are present.

Mines. Five Star, Golden Arrow, Huff, Mayflower, New Hope, Santa Maria.

Jacksonville

Location. This district is in southwestern Tuolumne County. It is in that portion of the Mother Lode belt that extends through the vicinity of southeast Jacksonville to the vicinity of Moccasin Creek. The Jamestown district is just to the north, the Big Oak Flat Groveland district is to the east, and the Coulterville district is to the southeast.

History. Jacksonville, named for Colonel Alden Jackson—as was the town of Jackson in Amador County—was founded as a supply center in 1848. The placer deposits here were extremely rich, credited with a production of $9 million. Lode mining began in the district in the late 1850s. The Eagle-Shawmut mine was operated on a large scale from 1897 until 1942. The district was served for some years by the Hetch Hetchy Railroad.

Geology and Ore Deposits. The deposits occur near or at the contact between serpentine on the west and slate and schist on the east interlayered with a number of narrow bands of greenstone. The ore deposits consist chiefly of large but low-grade bodies of pyrite ankerite-quartz and mariposite-ankerite-quartz rock and numerous pyrite-quartz stringers. Most of the gold values are in the sulfides; there is not much free gold. The ore zones were as much as 180 feet in thickness. There are several adjacent but relatively barren massive bull quartz veins. Mining extended to depths of 3000 feet. The ore usually contained 1/7 ounce or less gold per ton; for years the mill heads at the Eagle-Shawmut mine averaged about $2.75 per ton.

Mines. Clio $100,000+, Eagle-Shawmut $7.4 million, Harriman $100,000+, Mammoth $100,000, Moccasin, Orcutt, Republican, Tarantula $100,000+, Wheeler.

Jamestown

Location. The Jamestown district is in western Tuolumne County. It consists of that portion of the Mother Lode belt that extends from French Flat southeast through Rawhide, Jamestown, Quartz Mountain, and the town of Stent to the vicinity of the Belcher mine, a distance of about eight miles. It also has been called the "Jimtown" district.

History. The streams and rich surface ores were first worked in the gold rush. Jamestown was established in 1848 by Colonel George F. James, a lawyer. Hydraulic mining began at Stent soon afterward, and the lode mines were active from the 1860s on. The placers at nearby Campo Seco yielded $5.5 million and those at Jamestown $3 million. From around 1890 to World War I lode mining was a major industry; in 1906 more than 300 stamps were "dropping" in the various mills. There was some activity again during the 1920s and appreciable activity during the 1930s. There has been minor prospecting and development work in recent years at a few of the mines. The value of the total output of this district is estimated at more than $30 million.

Geology. In the north portion of the district, the deposits occur along a northwest-striking contact with serpentine to the southwest and phyllite, slate, and metaconglomerate to the northeast. In the central and south portion, the deposits are at or near the contact between massive greenstones and slates on the west and chlorite and amphibolite schist to the east. Latite of Tuolumne Table Mountain crosses the belt north of Jamestown, and Tertiary gravel deposits underlie the latite in the vicinity of the town of Rawhide and to the southwest. At Quartz Mountain the Mother Lode belt swings from a northwest-southeast strike to almost due south.

Ore Deposits. Outcrops consist of massive quartz veins up to several tens of feet in thickness, adjacent bodies of ankerite-quartz-mariposite rock which sometimes are scores of feet thick, as well as bodies of mineralized schist and numerous parallel quartz stringers. These deposits often contain abundant disseminated sulfides (as much as eight to 10 percent of the total rock), which are mostly pyrite. The gold occurs in the native state or with pyrite. Milling-grade ore usually averaged 1/7 to 1/3 ounce gold per ton, but the ore shoots were large. The ore shoots had stoping lengths of as much as 400 feet or more, and several veins were mined to inclined depths of several thousand feet. A number of high-grade pockets have been found in this district. In places silver is abundant, and tellurides have been encountered.

Mines. Alabama $150,000, Alameda, Anderson, App-Heslep $6.5 million, Belcher, Crystalline $100,000, Defender, Dutch-Sweeney $3 million, Erin-go-bragh $282,000, Golden Rule, Harvard $2 million to $3 million, Hitchcock, Jumper $5 million, Mazeppa, New Era, Nugget, Omega, Rappahannock, Rawhide $6 million, Santa Ysabel $1.5 million.

Sonora

History. This is one of the famous pocket-mining districts of the Sierra Nevada east gold belt. Sonora, the seat of Tuolumne County, was founded in 1848 soon after the discovery of rich placer deposits here and at Shaws Flat to the north. It was named for the state of Sonora in Mexico. The placers were extremely rich; the Sonora placers were credited with an output of $11 million and those of Shaws Flat yielded $6 million. The famous Holden Chispa nugget, which weighed over 28 pounds, was taken from within the city limits of Sonora in Holden's Gardens. In 1879 the Bonanza mine, also in town, yielded a pocket that contained $300,000. Later, large amounts of beautifully crystallized gold with tellurides were recovered from the Sugarman and Negro mine. Pocket mining continued almost steadily until World War II, and there has been some prospecting and development work since.

Geology. The central part of the Sonora district is underlain by a belt of crystalline limestone, which extends south from Columbia. The limestone is associated with slate, schist, and quartzite; all are part of the Calaveras Formation (Carboniferous to Permian). To the west is amphibolite and to the east is granodiorite.

Ore Deposits. The gold-quartz veins are largely confined to the slate, schist, and amphibolite. Much of the output from the lode mines has been from small but extremely rich pockets. The high-grade ore commonly contains crystallized gold, and in a few places the telluride minerals petzite and sylvanite are present. The veins, usually only a few feet wide, are often associated with diorite and aplite dikes. The rich early day placer deposits were in deep crevices and potholes in the limestone.

Mines. Aetna, Bonanza $1.5 million, Eureka, Fairview, Gerrymander, Golden Gate $100,000+, Hope $200,000+, Josephine, Lazar $100,000+, Lewis, Manzanita, O'Hara $100,000+, Rainbow, San Guiseppe, Sell, Stockton, Stuart and Morris, Sugarman $1 million, Tanzy, Vandelier.

Soulsbyville

Location. The Soulsbyville district is in west-central Tuolumne County in the general vicinity of the towns of Soulsbyville and Tuolumne. It includes the Arrastraville and Buchanan areas.

History. This district was placer-mined during the gold rush. Lode mining began in the early 1850s, and there was a rush to the district that began in 1858, after Ben Soulsby discovered rich ores. The mines were worked on a major scale until about 1915. There was some activity during the 1920s and 1930s, and there has been minor prospecting and development in a few of the mines since. This has been the most productive district in the Sierra Nevada east gold belt, with a total output value to be at least $20 million.

Geology. Granitic rocks, of which granodiorite predominates, underlie the west-central portion of the district. These granitic rocks are intrusive into slate, schist, phyllite, and quartzite of the Calaveras Formation (Carboniferous to Permian). Limestone is to the south and west, and the interstream ridges to the north are capped by andesite. Numerous dioritic and aplitic dikes are present, often associated with gold-quartz veins.

Ore Deposits. A large number of unoriented goldquartz veins occur in both the granitic and metamorphic rocks, usually ranging from one to five feet in thickness, The ore bodies are often lenticular in shape and contain native gold and often abundant sulfides, especially galena, which is nearly always associated with the gold. Milling-grade ore usually ranged from 1/2 to one ounce of gold per ton, and considerable high-grade ore has been mined in the district. 'The maximum depth of development is 1500 feet.

Mines. Agua, Caliente, Black Oak $3.5 million, Black Hawk, Blue Lead, Buchanan $600,000, Carlotta, Chickenfeed, Columbus $100,000+, Consolidated Eureka, Dead Horse, Draper $1 million, Dreisen, Empire, Fair Maiden, Fair Oaks, Garfield, Gilson (Platt & Gilson) $1.25 million, Grizzly $1.5 million, Hattie Ester, Hunter $300,000, Jigger Bill Brothers, Junction, Lady Washington, Laura & North Star, Louisiana, Mammon, New Albany, Ophir, Park and Mason, Phoenix, Providence $700,000, Prudhomme, Seminole, Spring Gulch $250,000, Soulsby $5.5 million, Starr King $100,000, South United $1.7 million, Waif, Wheal Ruff, Worcestor.

Tuttletown

Location. This district is in the Mother Lode belt in northwestern Tuolumne County. The Carson Hill district lies to the north and the Jamestown-Rawhide district to the south. It includes the Jackass Hill and French Gulch areas.

History. The streams and rich surface placers were worked during the gold rush. The area was known as Mormon Gulch in 1848 but was renamed for Judge Anson H. Tuttle, who built the first log cabin there. The Gillis brothers came here from Virginia City and it was in their cabin that Mark Twain stayed for five months in 1864-65 and Bret Harte for a night. A replica of the Gillis cabin is now a tourist attraction. This area became a well-known pocket-mining district during the 1860s and was mined almost continuously until World War II. Some work has been done at the Gross and Street mine since then.

Geology and Ore Deposits. A northwest-trending belt of amphibolite traverses the central portion of the district. Phyllite and slate are to both the northeast and southwest and serpentine lies to the south. Although this district lies between the Carson Hill and Jamestown districts, where large ore bodies were mined, most of the output here has been from small, rich pockets. These pockets occur in quartz veins and stringers and contain native gold, abundant pyrite, galena, and tellurides. The country rock adjacent to the veins commonly contains disseminated pyrite and ankerite.

Mines. Albion Cons., Alta, Anti-Chinese, Athena, Ball, Bown, Cardinelle, Chileno, Gagnere, Gross and Street, Marryatt, Norwegian $200,000+, Patterson, Tarantula, Toledo.

YUBA COUNTY

• Since the commencement of gold dredging operations in 1904, at Hammonton and Marigold, on the Yuba River, the gold output of the county has steadily increased until Yuba is now one of the most important gold producing counties in the state.

In 1904 there was one company operating two dredges during the months of August to December, in 1905 there were eight dredges in operation during part of the year, and in 1906 there were two dredging companies operating part of the year, the Marysville Dredging Company with 2 and the Yuba Consolidated Gold Fields with 8 dredges. During 1907, 12 dredges operated practically the entire year, and during 1908 there were 14 dredges in operation part of the year.

In 1909 the Marysville Dredging Company constructed and put in operation one dredge and dismantled another, so that there were the same number of dredges in active operation as in 1908.

Dredging in the Yuba River district began in 1904 when W. P. Hammon of San Francisco and the late R. D. Evans of Boston, Mass., commenced operations in August of that year with two California type dredges, the Yuba No. 1 and No. 2.

Previous to the building of these dredges Messrs. Hammon and Evans spent about two years in examining and acquiring the ground located in and adjacent to the Yuba Basin. During this time about 300 test holes were sunk with drills, to depths of from 60 to 70 feet and nearly $100,000 was expended in the work. The ground prospected extended over an area nearly five miles long with an average width of about one mile. The widest part of the property is located about four miles from where the Yuba River leaves a narrow canyon in breaking from the foothills of the Sierra Nevada, and spreads over a wide flat area about two miles across, known as the Yuba Basin.

The depth of the ground is due to the extensive hydraulic mining operations which for years were carried on along the headwaters and tributary streams of Yuba River. It is estimated that aside from the natural sedimentation due to erosion, that nearly half a billion cubic yards of hydraulic tailings have been carried down yearly by the flood waters and deposited in the river valley. The hydraulic tailings in the Yuba Basin range in depth from 10 feet to 40 feet and overlie the gold-bearing river gravel. These tailings carry some gold, but not enough to pay for working.

The old river gravel underlying the hydraulic tailings rests on a volcanic ash bedrock, similar in character to that of the Oroville district. In some places drill holes showed gravel and some gold below this bedrock, 90 feet and 110 feet from the surface, but usually the volcanic ash lies on the true bedrock.

The gold sometimes forms several pay-streaks in the old river gravel but is seldom found to extend to the volcanic ash bedrock. The ground yields from 10 cents to 30 cent per cubic yard from top to bottom. In dredging part of the bedrock, which is soft and sticky like the false bedrock in other districts, is usually dug. The natural water level ranges from several feet above to about four feet below the surface of the ground.

In 1905 the Yuba Consolidated Gold Fields was organized to take over the holdings acquired by W. P. Hammon and R. D. Evans. This company was incorporated under the laws of the State of Maine, with a capitalization of $12,500,000. The officers are: President, Geo. L. Huntress; secretary and treasurer, R. E. Paine, 50 Congress street, Boston, Mass.; managing director, W. P. Hammon; general manager, Newton Cleaveland, Marysville, Cal.; superintendent Geo. J. Carr, Hammonton, Cal. The holdings of the company comprise an area of 3,000 acres.

The Marysville Dredging Company began operations in 1905 under the direction of R. E. Cranston, Sacramento, Cal. The officers of the company are: Directors, F. Lothrop, president; Q. A. Shaw, vice-president; Alexander Agassiz; T. L. Livermore; H. L. Higginson; R. L. Agassiz, treasurer, 124 Ashburton Place, Boston, Mass.; A. D. Snodgrass, cashier; A. B. Strock, superintendent, Marigold, Cal.; Bukeley Wells, general manager, Marysville, Cal.

The holdings of the company adjoin those of the Yuba Consolidated Gold Fields on the west and comprise an area of about 1,200 acres located in section 36, township 16 north, range 4 east, and sections 1 and 2, township 15 north, range 4 east along the Yuba River, about 6 miles northeast of Marysville. Of the total acreage about 600 acres are dredging ground, averaging in depth about 60 feet. The gravel and bedrock is the same in character as that of the Yuba Consolidated Gold Fields. This company is also cooperating with the Debris Commission in their work. [2]

Brown's Valley

Location. Brown's Valley is in north-central Yuba County about 15 miles northeast of Marysville. The Hammonton dredge field is just to the south. The area was mined during the gold rush, when very coarse placer gold was recovered. It was named for a prospector who reportedly recovered $12,000 from a quartz vein in a few weeks in 1850. Much lode mining was done here in the 1860s and 1870s, and intermittent activity continued through World War I. The Dannebroge mine was reopened in 1945 and has been intermittently active since. The lode mines are estimated to have yielded $3 million to $5 million.

Geology. The central portion of the district is underlain by a northwest-striking belt of fine-grained greenstone that is classified as metadiabase and andesite porphyry. Amphibolite lies to the east, valley alluvium to the west, and gravels of Yuba River to the south.

Ore deposits. A number of nearly west-striking quartz veins in greenstone dip either north or south. The veins range from one to 10 feet thick. The ore contains free gold, pyrite, some galena, and chalcopyrite. Much of the recovered values have been free gold, but the sulfide concentrates often run high. Stoping lengths average around 150 feet, and the veins have been mined to inclined depths of 1700 feet.

Mines. Cleveland, Dannebroge, Hibbert and Burris, Jefferson, Pennsylvania, Too Handy.

Brownsville

Location. The Brownsville district is in northeastern Yuba County about 35 miles northeast of Marysville and 27 miles southeast of Oroville. It includes the lode-gold deposits here and in the Hansonville-Rackerby area, and the New York Flat placer "diggings". The town was named for I. E. Brown, who established a sawmill here in 1851.

Geology and ore deposits. The chief rocks in the area are greenstone with several gabbrodiorite intrusions. Amphibolite and some slate lie to the east. A number of north- and a few west-striking quartz veins commonly occur near intrusive-metamorphic contacts. The veins are one to seven feet thick and contain free gold with varying amounts of pyrite and other sulfides. The ore shoots are limited in size. Milling ore usually yielded 1/7 to 1/3 ounce of gold per ton, but some high-grade pockets were encountered.

Mines. Abbott, Arbucco, B. A. C., Beaver, Beehive (Mt. Hope) $100,000, Easy Money,

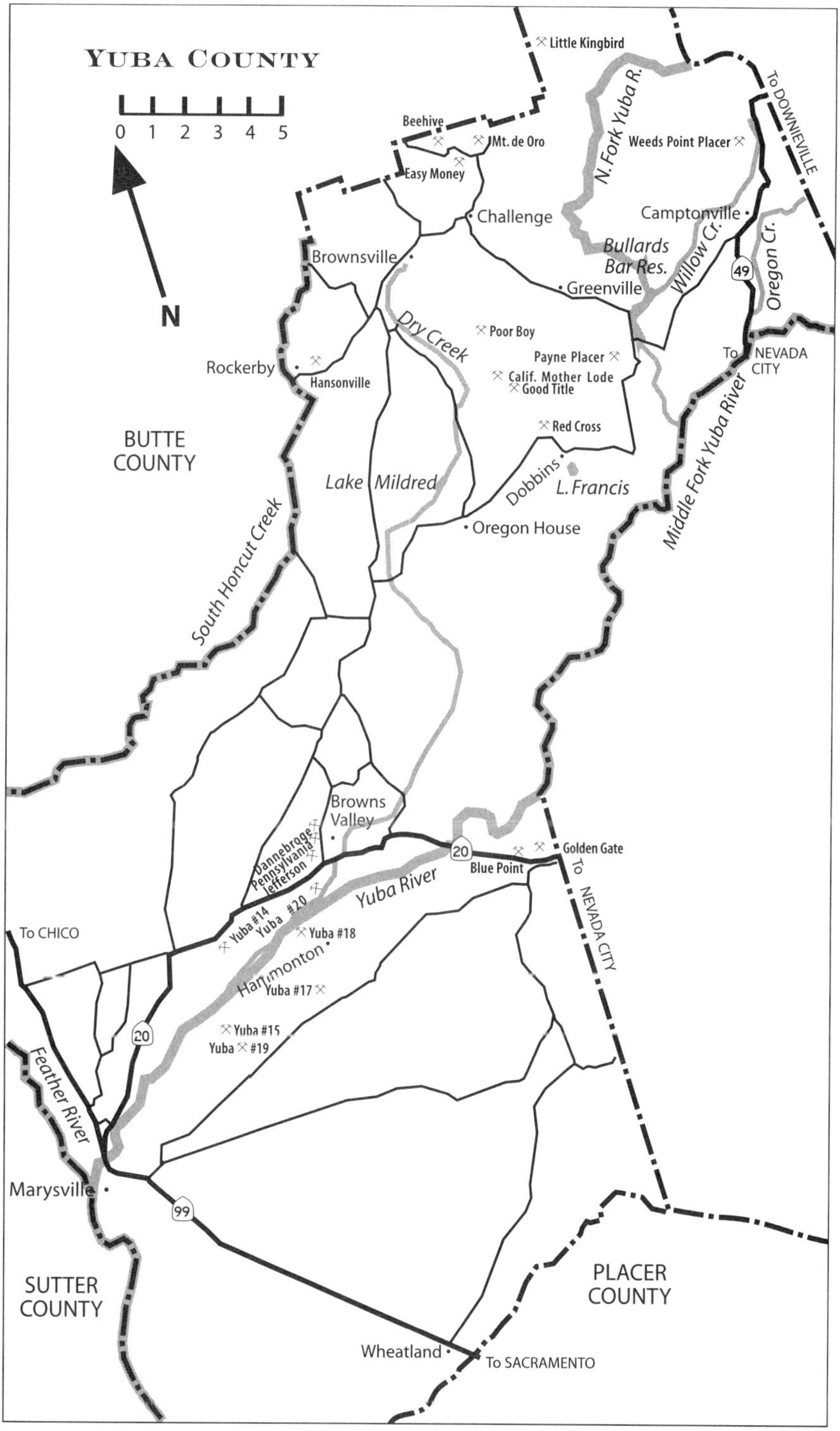

YUBA COUNTY
0 1 2 3 4 5
N
Little Kingbird
Beehive
Mt. de Oro
Easy Money
Weeds Point Placer
N. Fork Yuba R.
To DOWNIEVILLE
Challenge
Camptonville
Brownsville
Bullards Bar Res.
Willow Cr.
Oregon Cr.
49
Greenville
Poor Boy
Dry Creek
Payne Placer
To NEVADA CITY
Rockerby
Hansonville
Calif. Mother Lode
Good Title
Red Cross
BUTTE COUNTY
Dobbins
L. Francis
Lake Mildred
Middle Fork Yuba River
Oregon House
South Honcut Creek
Browns Valley
Dannebroge
Pennsylvania
Jefferson
20
Golden Gate
Blue Point
To NEVADA CITY
Yuba River
Yuba #14
Yuba #20
To CHICO
Yuba #18
Hammonton
Yuba #17
Yuba #15
Yuba #19
Feather River
Marysville
99
SUTTER COUNTY
PLACER COUNTY
Wheatland
To SACRAMENTO

Golden Key, Horseshoe, Manzanita, Napa and Oro, Ora Lewa, R. C. Rogers, Seaborg and Davis, Spanish, Twentieth Century Wonder, William Arthur.

Camptonville

Camptonville, in northeast Yuba County and western Sierra County, adjoins the Pike district on the east. Rich gold discoveries were made here in 1850-51. The town was named in 1854 for Robert Campton, the local blacksmith. The Pelton wheel, long used in mining machinery and electric generators, was invented here by Lester Pelton. The old town is now a tourist attraction. A number of moderate-sized deposits of Tertiary channel gravel were mined by hydraulicking. These include the deposits at Young's Hill, Weed's Point, and Galena Hill, They are part of the Tertiary Yuba River that extended southwest from Indian Hill to this district and then to North San Juan. Bedrock is amphibolite, slate, greenstone, and granodiorite.

Hammonton

Location. The Hammonton district is in south-central Yuba County along the lower Yuba River about 10 miles east of Marysville. It is a major dredge field that extends along the river about eight miles. It also is known as the Yuba River district.

History. The river and streams here were first worked during the gold rush by small-scale placer methods. However, this soon ceased because the river level was raised by a large influx of hydraulic mine tailings. Bucket-line dredging began in the district in 1903 under the direction of W. P. Hammon. In 1905 his interests were taken over by Yuba Consolidated Gold Fields, which had just been organized. This concern perfected large-scale bucket-line dredging here into one of the most efficient methods for mining placer gold. Yuba Dredge No. 20 was one of the largest gold dredges in existence. The district was dredged almost continuously from 1903 to 1968 and was the principal source of gold in California for some time. The estimated total output from dredging was estimated in 1964 at 4.8 million ounces.

However, operations have been gradually curtailed; in 1967 only two dredges were operating. On October 1, 1968 the last dredge was shut down, thus ending a major industry that had existed for nearly 70 years. More than a billion cubic yards of gold-bearing gravels were dredged. The extensive piles of gravel have become increasingly important as sources of aggregate.

Gold-Bearing Gravels. The gold-bearing gravels are in and south of the Yuba River, which flows west southwest through the area. Digging depths range from 60 to 80 feet on the upper end to 100 to 125 feet in the vicinity of the town of Hammonton. As much as 45 feet of the upper gravels are hydraulic mine tailings. Bedrock in the upper eastern end of the field consists of metamorphic rocks, while, in the central and western portions, the gravels are underlain by clay. The gold recoveries have been as follows: 12¢/yd in 1915-16; 14¢-15¢/yd in 1920-22; 8¢-9¢/yd in 1928-29; 12¢/yd in 1948-49, and 16.56¢/yd in 1959. The hydraulic tailings were reported to have averaged 6¢/yd at the old price. The gravels are medium to fine and are free-washing. Minor amounts of platinum were recovered.

According to the April, 1960 issue of *Mining World,* in 1959 Yuba Cons. reported four dredges treated 16,642,265 cu. yds. with an average content of 16.56¢ per yard. Reserves in the area were estimated to be about 93 million cu. yds. Estimates are that about 235 million yds of gold-bearing gravels are in the field but beyond depths of existing equipment.

Operations. Hammon and Evans, 1903-05 (bought by Yuba Cons.), two dredges; Marysville

Dredging Co., 1906-25 (bought by Yuba Cons.), five dredges; Pacific Gold Dredging Co., 1916-23, one dredge; Yuba Cons. Goldfields, 1905-1968, 21 dredges, not all worked at the same time.

Smartsville

Location. This district is located in western Nevada and eastern Yuba Counties 20 miles east of Marysville and 15 miles west of Grass Valley. It includes the Mooney Flat, Sicard Flat and Timbuctoo areas. It is mainly a placer-mining district.

History. The streams were placer-mined during the gold rush. The town was named for James Smart, who built a hotel there in 1856. The area was extensively hydraulicked from around 1855 to 1877. Some drift mining also was done during these years and continued through the early 1900s, but little work has been done here since. The value of the total output of the district is unknown. In 1877, it was reported to have been $13 million.

Geology. The main channel of the Tertiary Yuba River enters the area from the north and goes through Mooney Flat. It then curves west and northwest through Smartsville and Timbuctoo and then west to Sicard Flat. The gravel deposits are extensive and up to 200 feet thick. The lowest "blue" gravel on bedrock was the richest. The average yield during the major early operations was 37 cents in gold per yard, but some drifting yielded up to $3 per yard at the old price of gold. It has been estimated that 46.5 million yards were removed prior to 1891. Bedrock is greenstone, and in places the gravel is overlain by andesite.

BIBLIOGRAPHY

Allen, W. W., and Avery, R. B., 1893, California gold book: San Francisco and Chicago.

Ailing, M. N., 1922, Ancient river-bed deposits in California: Engineering and Min. Jour., vol. 1, Sept. 1922, pp. 134-140, and Oct. 1922, pp. 161-166.

Aubury, L. E., Winston, W. B., and Janin, Charles, 1910, Gold dredging in California: California Min. Bur. Bull. 57, 305 pp.

Bell, James E., 1956, Gold: U.S. Bur. Mines Bull. 556, pp. 315-326.

Bowie, A. J., Jr., 1905, A practical treatise an hydraulic mining in California: D. Van Nostrand Co., New York, 313 pp.

Browne, J. Ross, 1868, Mineral resources of the states and territories west of the Rocky Mountains: U.S. Government.

Browne, J. Ross, and Taylor, J. W., 1867, Mineral resources of the states and territories west of the Rocky Mountains: U.S. Government.

California Miners' Association, 1899, California's mines and minerals, San Francisco, 445 pp.

Clark, Lorin D., 1960, Foothills fault system, western Sierra Nevada: Geol. Soc. America Bull., vol. 71, pp. 483-496.

Clark, William B., 1957, Gold: in California Div. Mines Bull. 176, pp. 215-226.

Cloos, Ernst, 1935, Mother Lode and Sierra Nevada batholith: Jour. Geology, vol. 43, pp. 225-249.

Del Mar, Alexander, 1902, A history of precious metals: New York, 464 pp.

Del Mar, Alexander, 1911, Gold nuggets of California: Min. & Sci. Press, vol. 102, P. 629.

Doolittle, J. E., 1905, Gold dredging in California: California Min. Bur. Bull. 36, 120 pp.

Dunn, R. L., 1888, Drift mining in California: California Min. Bur. Rept. 8, pp. 736-770.

Dunn, R. L., 1890, River mining: California Min. Bur. Rept. 9, pp. 262-281.

Dunn, R. L., 1894, Auriferous conglomerate in California: California Min. Bur. Rept. 12, pp. 459-471.

Edman, J. A., 1907, Auriferous black sands of California; California Min. Bur. Bull. 45, 19 pp.

Emmons, W. H., 1937, Gold deposits of the world: McGraw-Hill Book Co., Inc., New York, 562 pp.

Fairbanks, H. W., 1890, Geology of the Mother Lode region: California Min. Bur. Rept. 10, pp. 22-90.

Gilbert, G. K., 1917, Hydraulic-mining debris in the Sierra Nevada: U.S. Geol. Survey Prof. Paper 105, 154 pp.

Goodwin, J. Grant, 1957, Lead and zinc in California: California Jour. Mines and Geology, vol. 53, pp. 353-724.

Goodyear, W. A., The auriferous gravels of California: Min. & Sci. Press, vol. 39, Sept. 20, 1879, pp. 182-183.

Haley, C. S., 1923, Gold placers of California: California Min. Bur. Bull. 92, 167 pp.

Hammond, John Hays, 1890, The auriferous gravels of California: California Min. Bur. Rept 9, pp. 105-138.

Hammond, John Hays, 1890, Mining of gold ores in California California Min. Bur. Rept. 10, pp. 852-882.

Hanks, H. G., 1882, Placer, hydraulic, and drift mining: California Min. Bur. Rept. 2, pp. 28-192.

Henderson, C. W., 1922, The history and influence of mining in the western United States. Ore deposits of the western states, Lindgren volume, Am. Inst. Min. Engrs., New York, pp. 730-784.

Hill, J. M., 1912, The mining districts of the western United States, with a geologic introduction by Waldemar Lindgren: U.S. Geol. Survey Bull. 507, pp. 17-20, 77-133.

Hill, J. M., 1915, Some mining districts in northeastern California and northwestern Nevada; U.S. Geol. Survey Bull. 594, pp. 133-141.

Hill, J. M., 1929, Historical summary of gold, silver, copper, lead, and zinc produced in California, 1848 to 1926: U.S. Bur. Mines Econ. Paper 3, 22 pp.

Hulin, C. D., 1933, Geological relations of ore deposits in California: Ore deposits of the western states, Lindgren volume, A.I.M.E., New York, pp. 240-253.

Janin, Charles, 1918, Gold dredging in the United States: U.S. Bur. Mines Bull. 127, 226 pp.

Jarman, Arthur, 1927, Report of the Hydraulic Mining Commission upon the feasibility of the resumption of hydraulic mining in California: California Min. Bur. Rept. 23, pp. 44-116.

Jenkins, Olaf P., 1935, New technique applicable to the study of placers: California Div. Mines Rept. 31, pp. 193-210.

Jenkins, Olaf P., et al, 1948, Geologic guidebook along Highway 49-Sierran gold belt the Mother Lode Country- California Div. Mines Bull. 141, 164 pp.

Joslin, G. A., 1945, Gold: California Div. Mines Bull. 130, pp. 122-151.

Knopf, Adolph, 1929, the Mother Lode system of California: U.S. Geol. Survey Prof. Paper 157, 88 pp.

Koschmann, A. H., and Bergendahl, M. H., 1962, Gold in the United States: U.S. Geological Survey Mineral Investigations Resources Map MR-24.

Koschmann, A. H., and Bergendahl, M. H., 1968, Principal gold producing districts of the United States: U.S. Geol. Survey Prof. Paper 610, pp. 53-84.

Lindgren, Waldemar, 1894c, Sacramento folio, California: U.S. Geological Survey Geol. Atlas of the U.S., folio 5, 3 pp.

Lindgren, Waldemar, 1894b, Marysville folio, California: U.S. Geol. Survey Geol. Atlas of the U.S., folio 17, 2 pp.

Lindgren, Waldemar, 1896c, Nevada City special folio, California: U.S. Geol. Survey Geol. Atlas of the U.S., folio 29, 7 pp.

Lindgren, Waldemar, 1896b, Pyramid Peak folio, California: U.S. Geol. Survey Geol. Atlas of the U.S., folio 31, 8 pp.

Lindgren, Waldemar, 1896c, Characteristic features of the California gold-quartz veins: Bull. Geol. Soc. America, pp. 221-240.

Lindgren, Waldemar, 1897, Truckee folio, California: U.S. Geol. Survey Geol. Atlas of the U.S., folio 39, 8 pp.

Lindgren, Waldemar, 1900, Colfax folio, California-. U.S. Geol. Survey Geol. Atlas of the U.S., folio 66, 10 pp.

Lindgren, Waldemar, 1909, Resources of the United States in gold, silver, copper, lead, and zinc: U.S. Geol. Survey, Bull. 394, pp. I 1 4-156.

Lindgren, Waldemar, 1912, The mining districts of the western United States: U.S. Geol. Survey Bull. 507, pp. 5-13, 17-20.

Lindgren, Waldemar, 1933, Gold-quartz veins of the Sierra Nevada: Mineral Deposits, McGraw-Hill Book Company, New York, pp. 616-627.

Lindgren, Waldemar, and Turner, H. W., 1894, Placerville folio, California: U.S. Geol. Survey Geol. Atlas of the U.S., folio 3, 3 pp.

Lindgren, Waldemar, and Turner, H. W., 1895, Smartsville folio, California: U.S. Geol. Survey Geol. Atlas of the U.S., folio 18, 6 pp.

Logan, C. A., 1919, Platinum and allied metals in California: California Min. Bur. Bull. 85, 120 pp.

Logan, C. A., 1935, Mother Lode gold belt of California: California Div. Mines Bull. 108, 240 pp.

Logan, C. A., 1950, Gold: California Div. Mines Bull. 156, pp. 503-514.

Merwin, Roland W., 1968, Gold resources in the Tertiary gravels of California: U.S. Bur. Mines Technical Progress Report, 14 pp.

Mining and Scientific Press, 1860-1921, various entries, vols. 1-125.

Moore, Lyman, 1968, Gold resources of the Mother Lode belt, California: U.S. Bur. Mines Technical Progress Report 5, Heavy Metals Program, 22 pp.

Paul, Rodman W., 1947, California gold, the beginning of mining in the far west, Harvard Univ. Press, Cambridge, 380 pp.

Peterson, D. W., Yeend, W. E., Oliver, H. W., and Mattick, R. E., 1968, Tertiary gold-bearing channel gravel in northern Nevada County, California: U.S. Geol. Survey Circular 566, 22 pp.

Preston, E. B., 1895, California gold mill practices: California Min. Bur. Bull. 6, 85 pp.

Ransome, F. L., 1900, Mother Lode district folio: U.S. Geol. Survey Geol. Atlas of the U.S., folio 63, 11 pp.

Raymond, R. W., 1869, 1870, 1871, 1872, 1874, 1875, and 1876, Mineral resources of the states and territories west of the Rocky Mountains: U.S. Government.

Reid, John A., The east country of the Mother Lode: Min. and Sci. Press, Mar. 2, 1907.

Rickard, Thomas A., 1932, A history of American Mining: D. Appleton-Century Co., Inc., New York and London, 419 pp.

Ridgeway, R. H., 1929, Summarized date of gold production: U.S. Bur. Mines, Econ. Paper 6, 63 pp.

Risdon Iron Works, 1885, Gold mines and mining in California: George Spaulding and Co., San Francisco, 349 pp.

Ryan, J. P., 1960, Gold: U.S. Bur. Mines Bull. 585, pp. 347-356.

Skidmore, W. A., 1885, Gold and silver mining in California, past, present, and prospective: Report of the Director of the Mint, pp. 525-557.

Storms, W. H., 1900, Mother Lode region of California: California Min. Bur. Bull. 18, 154 pp.

Turner, H. W., 1894a, Jackson folio, California: U.S. Geol. Survey Geol. Atlas of the U.S., folio 11, 6 pp.

Turner, H. W., 1894b, Notes on the gold ores of California: Am. Jour. Sci., 3rd ser., 47.

Turner, H. W., 1897, Downieville folio, California: U.S. Geol. Survey Geol. Atlas of the U.S., folio 37, 8 pp.

Turner, H. W., 1898, Bidwell Bar folio, California: U.S. Geol. Survey Geol. Atlas of the U.S., folio 43, 6 pp.

Turner, H. W., 1899, Replacement deposits in the Sierra Nevada: Jour. Geol. pp. 389-400.

Turner, M. W., 1905, Notes on the gold ores of California: Am. Jour. Sci., 3rd ser., 49.

Turner, H. W., and Ransome, F. L., 1897, Sonora folio, California: U.S. Geol. Survey Geol. Atlas of the U.S., folio 41, 7 pp.

Turner, H. W., and Ransome, F. L., 1898, Big Trees folio, California: U.S. Geol. Survey Geol. Atlas of the U.S., folio 51, 8 pp.

Weatherbe, D'Arcy, 1907, Dredging for gold in California: Mining and Scientific Press, Son Francisco, 217 pp.

Whitney, J. D., 1875, The auriferous gravels of the Sierra Nevada: Mem. Mus. Comp. Zool., Harvard College, vol. 6, 569 pp.

Yale, C. G., 1896, Total production of gold in California since 1848 by years according to different authorities: California Min. Bur. Rept. 13, pp. 64-65.

APPENDIX A: LEGISLATION

CALIFORNIA PUBLIC RESOURCES CODE

CHAPTER 6.5: PLACER MINING WATER POLLUTION

2551. *Exclusions.* This chapter does not apply to any placer mine operator who holds a permit to operate from the California Debris Commission.

2552. *Recordation of verified statement.* Any person, firm, or corporation who engages in the operation of a placer mine on any stream or on the watershed of any stream tributary directly or indirectly to the Sacramento or San Joaquin rivers shall record a verified statement in the office of the county recorder of the county in which his mine is situated, within 30 days after the commencement of operations.

2553. *Contents of verified statement.* The verified statement shall be verified by the operator or by some one in his behalf and shall contain the following information:

(a) A description of the ground proposed to be mined by placer mining methods, described by United States Government subdivisions where possible.

(b) The names and addresses of the owners of the ground.

(c) The names and addresses of the operators of the mine.

(d) The proposed means or method of placer mining operation.

(e) The means which the operator proposes to use to prevent the pollution of any stream by the effluent from such operations.

SUGGESTED FORM

Verified Declaration of Operator of Placer Mine

I, *[name]* declare under penalty of perjury:

1. That *[name]* proposes to mine the following ground by placer mining methods:*[describe by United States Government subdivisions where possible]*.

2. That the names and addresses of the owners of said ground are as follows: *[insert names]*.

3. That the names and addresses of the owners of said mine are as follows:*[insert names]*.

4. That the proposed means or method of placer mining operation are as follows: *[insert]*.

5. That the following means will be used to prevent pollution of any stream from said operation: *[describe]*.

Dated: *[date]*, at *[place]*

[Signature of Declarant]

2554. *Notice of change of address or ownership.* Within 10 days after an owner or operator changes his address, or transfers the ownership or changes the operator of any such mining property, a notice setting forth the names and addresses of the new owners or operators shall be filed in the office of the county recorder.

2555. *Precautions to prevent pollution.* No placer mining operator shall mine by the placer process on any stream or on the watershed of any stream tributary directly or indirectly to the Sacramento or San Joaquin rivers without taking the following precautions to prevent pollution of the stream by the effluent from operations:

(a) Constructing a settling pond or ponds of sufficient size to permit the clarification of water used in the mining processes before the water is discharged into the stream.

(b) Mixing with the effluent from mining operations aluminum sulphate and lime, or an equivalent clarifying substance which will cause the solid material in the effluent to coagulate and thus avoid rendering the water in the stream unfit for domestic water supply purposes.

2556. *Transportation of dredger across stream.* Notwithstanding the provisions of subdivision (b) of Section 2555, any placer miner who is operating by dredging process and who desires to transport his dredger across a stream may conduct the dredger across the stream without constructing a settling pond, if he is of the opinion that the expense of constructing settling ponds in the stream would be unduly heavy.

Before transporting his dredger, the operator shall use the following procedure:

(a) He shall give a notice of his intent to cross a stream to the clerk or the secretary of each city or district owning or operating a domestic water supply the clarity of which is likely to be affected by the crossing operation.

(b) The notice shall be given at least seven days in advance of the date that he expects to cross the stream with his dredger.

(c) Upon the expiration of the notice the operator may during the following 48 hours conduct his dredger across the stream even though some turbidity may be caused by the crossing operation.

(d) After crossing the stream, the operator shall comply with the provisions of subdivision (b) of Section 2555.

2557. *Violation of chapter.* Any person, firm, or corporation who violates this chapter is guilty of a misdemeanor.

2558. *Operations constituting public nuisance.* The operation of any placer mine on ground not covered by a permit to the operator from the California Debris Commission, without compliance with the provisions of this chapter, is declared to be a public nuisance which may be enjoined upon suit brought by the district attorney of the county in which the operation has been conducted, or by any city or district whose domestic water supply is rendered unfit or dangerous for human consumption by the acts, or failure to act, of the operator. The superior court of the county in which the operation is conducted has jurisdiction to hear and determine the action and to award such relief as may be proper.

2559. *Existing remedies preserved.* Nothing in this chapter shall be construed to deprive the State, any city, county, district, person, firm, or corporation of any right to maintain any action or proceeding, in any jurisdiction, which it was entitled to maintain prior to the enactment of this chapter, or to receive or obtain in any such action any remedy accorded to it under existing law.

2601. *Equipment deemed fixtures.* Sluiceboxes, flumes, hose, pipes, railway tracks, cars, blacksmith shops, mills, and all other machinery or tools used in working or developing a mine, are to be deemed affixed to the mine.

2602. *Right to conduct hydraulic mining.* The business of hydrualic mining may be carried on within the State wherever and whenever it can be carried on without material injury to navigable streams, or the lands adjacent thereto.

2603. *Hydraulic mining defined.* Hydraulic mining as used in section 2602, is mining by means of the application of water, under pressure, through a nozzle, against a natural bank.

CALIFORNIA FISH AND GAME CODE

5653. *Use of vacuum or suction dredge equipment.*

(a) Before any person uses any vacuum or suction dredge equipment in any river, stream or lake of this state, the person shall submit an application for a permit for a dredge to the department, specifying the type and size of equipment to be used and other information as the department may require.

(b) The department may designate waters or areas wherein vacuum or suction dredges may be used pursuant to a permit, waters or areas closed to those dredges, the maximum size of those dredges which may be used, and the time of year when those dredges may be used. If the department determines that the operation will not be deleterious to fish, it shall issue a permit to the applicant. If any person operates any equipment other than that authorized by the permit or conducts the operation in any waters or area or at any time which is not authorized by the permit, or if any person conducts the operation without securing the permit, the person is guilty of a misdemeanor.

(c) The department shall issue a permit upon the payment, in the case of a resident, of a base fee of twenty-five dollars ($25), as adjusted under Section 713, when an onsite investigation of the project size is not deemed necessary by the department, and a base fee of one hundred thirty dollars ($130), as adjusted under Section 713, when the department deems an onsite investigation is necessary. In the case of a nonresident, the base fee shall be one hundred dollars ($100), as adjusted under Section 713, when an onsite investigation is not deemed necessary, and a base fee of two hundred twenty dollars ($220), as adjusted under Section 713, when an onsite investigation is deemed necessary.

(d) It is unlawful to possess a vacuum or suction dredge in areas, or in or within 100 yards of waters, which are closed to the use of vacuum or suction dredges.

SUGGESTED FORM

Petition for Writ of Mandate To Compel State Departnent of Fish and Game To Issue Permit for Dredge

[Title of Court and Cause]

1. Petitioner resides at *[address]*, City of *[city]* State of California.

2. Respondent is the Department of Fish and Game of the State of California and as such is the duly constituted authority to administer the provisions of the Fish and Game Code of the State of California relating to permits for use of vacuum or suction dredge equipment in any waters of the State of California.

3. On *[date]*, petitioner filed an application with respondent for a permit to *[specify type of equpment]* at *[describe location]* in the *[specify river, stream or lake]*, for the purpose of *[specify]*.

4. At all times mentioned in this petition, petitioner was fully qualified to be issued the above permit in the State

of California, inasmuch as petitioner *[state all qualifications outlined in the Fish and Game Code and pertinent regulations]*.

5. By reason thereof, petitioner is entitled to be issued the above permit in the State of California, and the refusal of respondent to so issue the permit constitutes an arbitrary and wrongful violation of its duties, in that petitioner has fully complied with all requirements for issuance of the permit and is, therefore, entitled under the provisions of the Fish and Game Code of the State of California to be issued the above permit.

6. Petitioner has performed all conditions precedent to the filing of this petition by *[insert all conditions precedent, including the exhaustion of administrative remedies]*.

7. Petitioner has made demand of respondent that petitioner be issued the above-mentioned permit, but the demand of petitioner has been refused and respondent does now refuse to issue a permit to petitioner.

8. At all times mentioned in this complaint, respondent has been able to issue a permit to petitioner, but notwithstanding this ability and in spite of the demand by petitioner that petitioner be issued a permit, respondent continues to refuse to so issue a permit.

9. Petitioner has no plain, speedy or adequate remedy in the ordinary course of law, other than the relief sought in this petition, in that this is the only proceeding in which petitioner may obtain the rights demanded in this petition.

Wherefore, petitioner requests that:

1. This court issue an alternative writ of mandate commanding respondent to issue the above mentioned permit to petitioner, or show cause before this court at a time specified by an order of the court why it has not done so and why a peremptory writ of mandate should not issue;

2. On the return of the alternative writ of mandate and the hearing of this petition, this court issue its peremptory writ of mandate commanding the respondent to do the things requested above;and

3. This court award petitioner costs of this proceeding and any further relief that to the court seems just and proper.

[Signature] [Verification]

5653.3. *Presentation of dredging equipment for inspection*. Any person required to possess a permit pursuant to Section 5653 shall present his or her dredging equipment for inspection upon request of a state or county fish and game warden.

5653.5. *"River, stream, or lake"*. For purposes of Section 5653, "river, stream, or lake" means the body of water at the current water level at the time of the dredging.

5653.7. *Closure of areas for unanticipated water level change*. In the event of an unanticipated water level change, when necessary to protect fish and wildlife resources, the department may close areas that were otherwise opened for dredging and for which permits were issued pursuant to Section 5653.

5653.9. *Adoption of regulations*. The department may adopt regulations to carry out Sections 5653, 5653.3, 5653.5, and 5653.7.

APPENDIX B: REGULATION

RESOURCES AGENCY OF CALIFORNIA DEPARTMENT OF FISH AND GAME SUCTION DREDGE REGULATIONS

The use of a suction dredge within the water-runs of California requires a Dredge Permit. Unless otherwise stated herein, the maximum dredge intake size permitted under a standard permit is 8 inches in diameter. Any dredge which has an intake nozzle or hose with an inside diameter that is larger than the permitted intake size, must have a constricting ring attached to the nozzle. This ring shall be of a solid one piece construction, with no openings other than the one intake opening of the allowable intake size. The constricting ring must be welded, or otherwise permanently attached, over the end of the intake nozzle; no quick release devices are permitted. The inner diameter of the intake hose may not be more than four inches larger than the permitted intake size.

AREAS FOR USE WITH A STANDARD PERMIT

Special Permits are not valid in these waters unless so specified in the Special Permit. A Special Permit must be obtained from the Department in order to operate a suction dredge in non-conformance with these regulations. Suction dredging into the bank of any stream is prohibited. Exceptions to these classifications are noted for each county.

CLASS A: Closed Waters. No dredging permitted at any time.
CLASS B: Open to dredging from July 1 through August 31.
CLASS C: Open to dredging from the fourth Saturday in May through October 15.
CLASS D: Open to dredging from from July 1 through September 15.
CLASS E: Open to dredging from July 1 through September 30.
CLASS F: Open to dredging from December 1 through June 30.
CLASS G: Open to dredging from the fourth Saturday in May through September 30.
CLASS H: Open to dredging throughout the year.

Amador County–East of Hwy. 49: Class C; elsewhere: Class H.

COSUMNES RIVER. From the Western Pacific Railroad bridge to the Latrobe Highway bridge: Class D; from the Latrobe Highway bridge to the confluence with the North and Middle Forks of the Cosumnes River: Class H. COSUMNES RIVER, SOUTH FORK. From the Middle Fork Cosumnes River to the county road bridge at River Pines: Class H. MOKELUMNE RIVER. From Burelia Road to Camanche Dam: Class A; from Camanche Dam to Pardee Dam: Class H; from Pardee Dam upstream: Class C. SUTTER CREEK. From Hwy. 49 to Pine Gulch Road: Class H.

Butte County–Class C.

BIG CHICO CREEK. From Manzanita Avenue in Chico to the head of Higgens Hole at location Sec. 31, T24N R3E: Class A. BUTTE CREEK. From Sutter County to the Durham-Oroville Highway bridge: Class H; from the Durham-Oroville Highway bridge to the intake of Centerville Ditch at location Sec. 10, T23N R3E: Class A. FEATHER RIVER. From Honcut Creek (Sec. 27, T17N R3E) to the Hwy. 70 bridge: Class B; from the Hwy. 70 bridge upstream to Oroville Dam: Class A.

Calaveras County–East of Hwy. 49: Class C; elsewhere: Class H.

CALAVERAS RIVER AND TRIBUTARIES. Below New Hogan Reservoir: Class B. MOKELUMNE RIVER. From Burelia Road to Camanche Dam: Class A; from Camanche Dam to Pardee Dam: Class H; from Pardee Dam upstream: Class C. STANISLAUS RIVER. From the San Joaquin River to Goodwin Dam: Class A; from New Melones Dam upstream, excluding Melones Reservoir: Class C.

El Dorado County–East of Hwy. 49: Class C; elsewhere: Class H.

AMERICAN RIVER, MIDDLE FORK. From its junction with the North Fork of the American River to the confluence with the Rubicon River: Class C. Recreational dredging is permitted in the Auburn State Recreational Area on an interim management basis. AMERICAN RIVER, SOUTH FORK. From Folsom Reservoir to the Hwy. 49 bridge at Coloma: Class C. AMERICAN RIVER, SOUTH FORK TRIBUTARIES. All tributaries from Folsom Reservoir upstream: Class C. COSUMNES RIVER. From the Western Pacific Railroad bridge to the Latrobe Highway bridge: Class D; from the Latrobe Highway bridge to the confluence with the North and Middle Forks of the Cosumnes River: Class H. COSUMNES RIVER, NORTH FORK. From the Middle Fork of the Cosumnes River to Somerset-Pleasant Valley Road bridge: Class H. COSUMNES RIVER, MIDDLE FORK. From the North Fork Cosumnes River to Bakers Ford on the Aukum-Somerset Road: Class H. COSUMNES RIVER, SOUTH FORK. From the Middle Fork

Cosumnes River to the county road bridge at River Pines: Class H. RUBICON RIVER AND TRIBUTARIES. Class C. Dredges with an intake larger than four inches are prohibited.

Fresno County–Within National Forest boundaries: Class C; elsewhere: Class H.

BIG CREEK AND TRIBUTARIES. Class A. COW CREEK AND TRIBUTARIES: Class A. DINKEY CREEK AND TRIBUTARIES: Class A. KINGS RIVER AND TRIBUTARIES. From Tulare Lake to Pine Flat Dam: Class A. SAN JOAQUIN RIVER. Upstream to Friant Dam (Millerton Lake): Class A. SYCAMORE CREEK AND TRIBUTARIES: Class A.

Kern County–Class H.

KERN RIVER AND TRIBUTARIES. From Isabella Dam upstream: Class A.

Lassen County–Class C.

PIT RIVER AND TRIBUTARIES: Class A.

Madera County–Within National Forest boundaries: Class C; elsewhere: Class H.

CHOWCHILLA RIVER. From Eastman Lake upstream to the West and East Forks of the Chowchilla River: Class A. CHOWCHILLA RIVER WEST FORK. From its mouth to Hwy. 49 bridge: Class A. EASTMAN LAKE: Class A. PORTUGUESE CREEK AND TRIBUTARIES: Class A. SAN JOAQUIN RIVER. Upstream to Friant Dam (Millerton Lake): Class A.

Mariposa County–Within National Forest boundaries: Class C; elsewhere: Class H.

CHOWCHILLA RIVER. From Eastman Lake upstream to the west and east forks of the Chowchilla River: Class A. CHOWCHILLA RIVER WEST FORK. From its mouth to Hwy. 49 bridge: Class A. EASTMAN LAKE: Class A. MERCED RIVER: Class C. MERCED RIVER, NORTH FORK: Class C. SAXON CREEK: Class A. SHERLOCK CREEK: Class A.

Merced County–Class H.

MERCED RIVER. From the San Joaquin River to Crocker-Huffman Dam, upstream from Snelling: Class A. SAN JOAQUIN RIVER. Upstream to Friant Dam (Millerton Lake): Class A.

Nevada County–East of Hwy. 49: Class C; elsewhere: Class H.

DEER CREEK. From Ponderosa Way below Rough and Ready Falls (Sec. 13, T16N R7E), to Hwy 49: Class C. Independence Creek And Tributaries. From Independence Lake upstream: Class A. MACKLIN CREEK. From its confluence with the Middle Fork Yuba River (Sec. 16, T19N R12E) upstream: Class A. WOLF CREEK. From the Tarr Ditch diversion (Sec. 10, T15N R8E) upstream: Class C.

Placer County–East of Hwy. 49: Class C; elsewhere: Class H.

AMERICAN RIVER, MIDDLE FORK. From its junction with the North Fork of the American River to the confluence with the Rubicon River: Class C. Recreational dredging is allowed in the Auburn State Recreational Area on an interim management basis. AMERICAN RIVER, NORTH FORK. From Folsom Reservoir to the Colfax-Iowa Hill Road bridge: Class C. From the Colfax-Iowa Hill Road bridge upstream to Heath Springs at location Sec. 26, T16N R14E: Class A. ANTELOPE CREEK AND TRIBUTARIES. Class B. AUBURN RAVINE AND TRIBUTARIES. Class B. BEAR RIVER. From Forty Mile Road to the South Sutter Irrigation District diversion dam: Class D. MINER'S RAVINE AND TRIBUTARIES. Class B. SECRET RAVINE AND TRIBUTARIES. Class B. RUBICON RIVER AND TRIBUTARIES. Class C. Dredges with an intake larger than four inches are prohibited.

Plumas County–Class C.

Sacramento County–Class H.

AMERICAN RIVER. From the Sacramento River to Nimbus Dam: Class A. COSUMNES RIVER. From the Western Pacific Railroad bridge to the Latrobe Highway bridge: Class D.

San Joaquin County–Class H.

Sierra County–Class C.

Independence Creek And Tributaries. From Independence Lake upstream: Class A. YUBA RIVER, NORTH FORK. From the Middle Fork of the Yuba River to Fiddle Creek: Class H.

Stanislaus County–Class H.

SAN JOAQUIN RIVER. Upstream to Friant Dam (Millerton Lake): Class A. STANISLAUS RIVER. From the San Joaquin River to Goodwin Dam: Class A; from New Melones Dam upstream, excluding Melones Reservoir: Class C. TUOLUMNE RIVER. From the Waterford bridge to La Grange Dam: Class A.

Tehama County–Class D.

Tuolumne County–East of Hwy. 49: Class C; elsewhere: Class H.

KAWEAH RIVER. Upstream of Kaweah Reservoir: Class A. CHERRY CREEK. Class B. CLAVEY RIVER. Class A. CURTIS CREEK. Class C. EAGLE CREEK. Class C. GRAPEVINE CREEK. Class B. HORTON CREEK. Class A. HUNTER CREEK. Class B. JAWBONE CREEK. Class B. KNIGHTS CREEK. Class C. MINNOW CREEK. Class A. POOR MAN CREEK AND TRIBUTARIES. Class A. ROSE CREEK. Class C. SIX-BIT CREEK AND TRIBUTARIES. Class A. STANISLAUS RIVER. From the San Joaquin River to Goodwin Dam: Class A; from New Melones Dam upstream, excluding Melones

Reservoir: Class C. SULLIVAN CREEK. Class C. TUOLUMNE RIVER, NORTH FORK, AND TRIBUTARIES. Class B. TURNBACK CREEK AND TRIBUTARIES. Class A. WOODS CREEK AND TRIBUTARIES. From Harvard Mine Road (Jamestown) downstream: Class C; from Harvard Mine Road upstream: Class A.

Yuba County–Class H.

CANYON CREEK. From its mouth to Sierra County, (Sec. 25, T20N R8E): Class C. YUBA RIVER. From Marysville to Hwy. 20: Class B; from Hwy. 20 to Englebright Dam: Class A.

RESTRICTED AREAS

Waters in national parks, national monuments, state parks and designated wilderness areas are closed by the managing agencies.

Parts of the Sequoia and Sierra National Forests—designated as the Kings River Special Management Area—are closed to dredging. Inquire at the appropriate U.S. Forest Service office.

The Auburn State Recreation Area restricts suction dredging; some open areas within the Auburn State Recreation Park Area are designated Recreation Gold Mining Free-use Sites.

FEDERAL WILD AND SCENIC RIVERS

Mining is restricted within 1/4-mile of these waters.

Kern River (Tulare County)–5,600 feet upstream of the Johnsondale bridge to the headwaters.

Merced River–From the upper end of Lake McClure at the 867-foot high-water level upstream to the headwaters.

Middle Fork Feather River–From Lake Oroville to Beckworth. Mining is prohibited in the wild section of the river. The U.S.F.S. has other regulations of mining claims within this section.

North Fork American River–From Auburn Reservoir to the Cedars. ALL NEW MINING IS PROHIBITED.

Tuolumne River–From Don Pedro Reservoir to Yosemite National Park.

APPENDIX C: ADDRESSES

RECREATIONAL GOLD PROSPECTING ORGANIZATIONS

Comstock Gold Prospectors–P.O. Box 20781, Reno NV 89515.

Gold Country Miners–(916) 725-4265; Dennis Ruff.

Gold Nugget Prospectors Club–9430 Mission, Riverside CA 92509.

Gold Prospectors Association of America–P.O. Box 891509, Temecula CA 92589-1509. (800) 551-9707, (909) 699-4749.

Golden Caribou Mining Club–P.O. Box 300, Belden CA 95915. (916) 283-0956; Steve Draper.

Motherlode Goldhounds–P.O. Box 4389, Auburn CA 95604.

Sacramento Detecting Buffs–4636 Nottingham Circle, Sacramento CA 95864.

Santa Clara Miners Association–P.O. Box 2861, Sunnyvale CA 94087-2861.

Santa Rosa Gold Diggers–(707) 544-5050, (707) 546-8460.

United Prospectors Inc.–7425 English Hills Rd., P.O. Box 95696-6329, Vacaville CA 95688-9528.

Valley Prospectors Club–(909) 883-7303; Dee Stapp.

Weekend Goldminers–P.O. Box 910, Dalonega, GA 30533.

West End Prospectors Club–P.O. Box 834, Fontana CA 92335.

Western Chapter FMDAC–5043 Strasbourg Way, Sacramento, CA 95842.

UNITED STATES FOREST SERVICE

Eldorado National Forest–100 Forni Road, Placerville, CA, 95667. (916) 644-6048.

Lassen National Forest–55 S. Sacramento St., Susanville, CA, 96130. (916) 257-2151.

Modoc National Forest–441 N. Main St., Alturas, CA, 96101. (916) 667-2247.

Plumas National Forest–P.O. Box 1500, Quincy, CA 95971. (916) 283-2050.

Sequoia National Forest–900 W. Grand Ave., Porterville, CA, 93257. (209) 784-1500.

Sierra National Forest–1130 "O" St., Fresno, CA, 93721. (209) 487-5155.

Stanislaus National Forest–19777 Greenley Road, Sonora, CA, 95370. (209) 532-3671.

Tahoe National Forest–Hwy. #49 & Coyote St., Nevada City, CA, 95959. (916) 265-4531.

UNITED STATES DEPARTMENT OF INTERIOR BUREAU OF LAND MANAGEMENT

California State Office–Federal Building, Room E-2841, Cottage Way, Sacramento, CA, 95825. (916) 978-4754.

Bakersfield District Office–800 Truxtun Ave., Room 311, Bakersfield, CA. 93301. (805) 861-4191.

Susanville District Office–P.O. Box 1090, 705 Hall St., Susanville, CA., 96103. (916) 257-5381.

Ukiah District Office–P.O. Box 1027, 555 Leslie St., Ukiah, CA., 95482. (707) 462-3873.

California Desert District Office–6221 Box Springs Blvd., Riverside, CA., 92507. (714) 697-5200.

STATE OF CALIFORNIA

State Of California Office Of Tourism–1121 "L" St., #103, Sacramento, CA, 95814.

California Dept. of Transportation–P.O. Box 942874, Room 2459, Sacramento, CA., 94274. (916) 654-3206.

California Dept. Of Parks & Recreation–P.O. Box 2390, Sacramento, CA, 95811. (916) 445-4624.

California Dept. Of Forestry–1416 Ninth St., Sacramento, CA, 95814. (916) 445-9920.

California Division Of Mines & Geology–P.O. Box 2980, Sacramento, CA, 95812. (916) 445-5716.

CALIFORNIA DEPARTMENT OF FISH & GAME

License & Revenue Branch–3211 S St., Sacramento. CA., 95816. (916) 653-4094.

Eureka–619 2nd. St., Eureka, CA., 95501. 707-445-6493.

Menlo Park–411 Burgess Dr., Menlo Park, CA., 94025. 415-688-6340.

Monterey–2201 Garden Rd., Montery, CA., 93490. 408-649-2870.

San Diego–1350 Front St., Room #2041, San Diego, CA., 92101. 619-237-7311.

Regional Offices–601 Locust St., Redding CA., 96001. (916) 225-2300; 1701 Nimbus Rd., Rancho Cordova, CA., 95670. (916) 355-0978; 7329 Silverado Trail, Napa, CA., 94558.; 1234 E. Shaw Ave., Fresno, CA., 93710. (209) 222-3761; 330 Golden Gate Shore #50, Long Beach, CA., 213-590-5132.